Biogeography

Ken Atkinson

Advanced
Topic*Master*

Series editor
Michael Raw

Acknowledgements

I would like to thank the colleagues who have provided photographs — Steve Carver, Pat Fargey, Michael Raw and Richard Smith. Thanks are due to friends who over the years have shared their biogeographical experience with me — Bill Archibold, Margaret Atherden, Bob Eyre, E. A. FitzPatrick and Richard Smith. Thanks also to the series editor, Michael Raw, for all his help and advice.

Philip Allan Updates, part of the Hodder Education Group, an Hachette Livre UK company, Market Place, Deddington, Oxfordshire OX15 0SE

Orders

Bookpoint Ltd, 130 Milton Park, Abingdon, Oxfordshire, OX14 4SB
tel: 01235 827720
fax: 01235 400454
e-mail: uk.orders@bookpoint.co.uk
Lines are open 9.00 a.m.–5.00 p.m., Monday to Saturday, with a 24-hour message answering service. You can also order through the Philip Allan Updates website: www.philipallan.co.uk

Printed in Spain

Philip Allan Updates' policy is to use papers that are natural, renewable and recyclable products and made from wood grown in sustainable forests. The logging and manufacturing processes are expected to conform to the environmental regulations of the country of origin.

Contents

Introduction

This book provides an up-to-date and detailed account of biogeography, which is part of the broader field of physical geography. It is primarily aimed at sixth-form students of AS/A-level geography, but the content of the book may be useful for first-year undergraduates. It may also be relevant for students of biology, ecology and environmental sciences.

The core subject matter of biogeography has traditionally been vegetation and soils. For much of the twentieth century, the focus was on natural vegetation, both worldwide and in small-scale studies of local plant communities. Soil studies were concerned mainly with describing soil sections in the field and measuring properties judged to be important in determining their fertility and food-producing capabilities.

In the twenty-first century, biogeography has become a dynamic subject, with concepts and techniques evolving continuously. Its content is being adapted in the face of environmental concerns. Subject matter has been added that is relevant to the challenges facing society through:

- declining biodiversity caused by the actual and predicted loss of plant and animal species
- the need for the conservation of wildland and wildlife in the face of increasing deforestation, destruction of wildlife habitat and loss of wilderness
- the impacts of, and responses to, environmental change
- the use of soils in a sustainable manner
- the restoration of damaged ecosystems

In short, the biogeographer of the twenty-first century will have an important role to play in helping to mitigate adverse impacts of society's demands on the Earth and its resources on the quality of vegetation, soils, wildland and wildlife.

The aim of this book is to provide information on the key biogeographical processes operating in the Earth's ecosystems. An understanding of basic physical, chemical and biological processes is fundamental to successful sustainable management. In addition to purely scientific study, the applied aspect of the subject will be needed to tackle environmental problems.

The content is in four parts, with two chapters in each. The first deals with the basic concepts of ecosystem structure and function and the distribution of plants and animals. The second part covers properties and formation of soils and the impacts on them of use by humans. The third takes a global view by

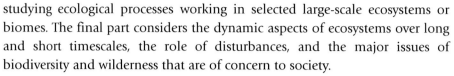

studying ecological processes working in selected large-scale ecosystems or biomes. The final part considers the dynamic aspects of ecosystems over long and short timescales, the role of disturbances, and the major issues of biodiversity and wilderness that are of concern to society.

As a student, you can use the book in several ways. The material reinforces knowledge and understanding of biogeography for essays, class discussions and other assignments. The figures, tables and photographs should be studied carefully. They have been chosen to illustrate and extend the concepts and facts covered by the text. They will improve understanding and could be used as a focus for class discussion and assignments. Today, magazines, television and the internet bombard us with images of the natural world from all corners of the Earth. However, the impact of the media can be transitory and detail is not retained in the mind. The illustrations in this book should provide a firmer and more permanent understanding. Redrafting the figures, photographs and tables in words and sketches is a good exercise to help focus understanding. Such sketches could be useful in assignments and examinations.

The activities in this book are devised to reinforce the scientific nature of biogeography by utilising a range of skills — graphical, hypothesis-testing, statistical and, most difficult of all, *thinking!*

Ken Atkinson

1 What is biogeography?

The nature of biogeography

The focus of biogeography has changed over time. In the nineteenth century, when the science of biology was advancing rapidly due to the discoveries of Charles Darwin, Joseph Hooker and Alfred Wallace, biogeography meant 'the study of the geography of plants, animals and other organisms; that is, their distribution over the Earth'. **Phytogeography** was concerned with the distribution of plants; **zoogeography** with the distribution of animals.

Past and present distributions of organisms are still studied by biogeographers, but today more attention is given to the relationships and interactions of organisms with their environment (other organisms, geology, soils, water and climate). This is the **ecological approach**. It requires some familiarity with the concepts of other subjects — geography, geology, soil science, hydrology, climatology and biology. Biogeography has also become analytical and quantitative, rather than being merely descriptive.

An applied science

Although it has theories and concepts, biogeography is an important applied science that seeks to improve the condition of human society. It has become concerned with:

- the **protection** and **conservation** of organisms, and the natural areas they inhabit
- **human impacts** on the environment
- debate on **sustainable use** and **environmental assessment**
- the modelling of **environmental change**, including the effects of changing climate, and the impacts of invading exotic species

Introductory definitions

Genera and species

Life on Earth consists of millions of organisms, ranging in size from the smallest virus to the largest blue whale. From Ancient Greek times, scientists have been concerned with **taxonomy**, i.e. naming and classifying this diversity of life. A **genus** (plural **genera**) is a group of organisms that are similar in appearance and structure, and are closely related in their evolution. A **species** is a group of organisms within a genus that can breed viably among themselves but not with other groups.

Naming organisms

In Britain there are two native oaks. Their names are based on the genus name *Quercus* plus the two species names, *robur* and *petraea*. The pedunculate oak of the lowlands is *Quercus robur*; the northern sessile oak found on higher ground in more northerly locations is *Quercus petraea*. Note three conventions about the scientific names of organisms:

- the use of Latin
- the use of italics
- the use of upper case for the initial letter of the name of the genus and lower case for that of the species

Common names, in English, are not italicised.
Names are usually descriptors of the organism:

- *Robur* is the Latin for 'hard' or 'strong', so *Quercus robur* means hard oak'.
- A species may be named after the discoverer — for example, *Pinus banksiana*, the North American jack pine, was named after the British botanist Joseph Banks.
- A species may be named after the place where it was discovered — for example, *Pinus halepensis*, the Aleppo pine, was named after the city of Aleppo in Syria.

Common names are used in this book, except in cases where a species does not have a common name. Field guides that identify and classify organisms such as birds, plants (**floras**) or marine mammals, normally give both scientific and common names.

Carl Linneaus

The system in use today for naming and classifying organisms was first devised by Carl Linneaus, an eighteenth century Swedish botanist. The use of Latin overcomes vagueness and misunderstanding that can arise from the use of common names, which usually differ from country to country. The system enables international collaboration. For example, a biogeographer in the UK can discuss vegetation with a biogeographer in Russia, despite the plants having different common names in different alphabets! Linneaus based his taxonomy on plant reproduction, appearance and structure. Like all systems of taxonomy, continual revision is necessary as concepts and techniques develop — for example, gene sequencing in biology.

Populations, communities, habitats and ecosystems

A **population** of an organism is a collection of plants or animals of the same species in a defined geographical area. It is assumed that the members of a population interact and interbreed frequently. For example, a population of humpback whales migrates in autumn from the west coast of North America to the Hawaiian Islands.

A biological **community** is composed of a number of populations of organisms. The heather moorland community in the British uplands and coral-reef communities in tropical oceans are examples.

A **habitat** is the environment or natural geographical area where a population is found. It is characterised by physical features such as soil conditions and moisture, and also by the other organisms in the area.

Ecosystems

The term **ecosystem** (Greek, *oikos*, a house) was coined by the biogeographer Sir Arthur Tansley in 1935. It includes the living biological populations, their physical and biological environment (Figure 1.1(a)), and the interactions that bind the populations and environment into one **biotic** complex (Figure 1.1(b)). Two of the interactions that play pivotal roles in ecosystems are the flow of energy (pp. 8–15) and the cycling of nutrients (pp. 16–22).

| Figure 1.1 | **Ecosystem components and interactions** |

(a) Components

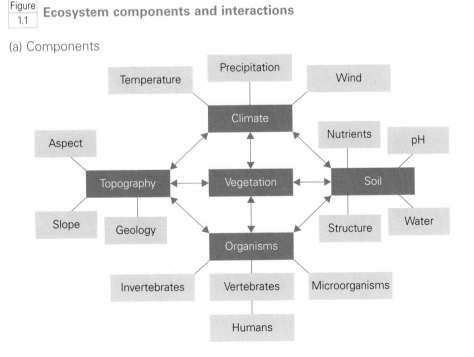

(b) Subsystems and interactions

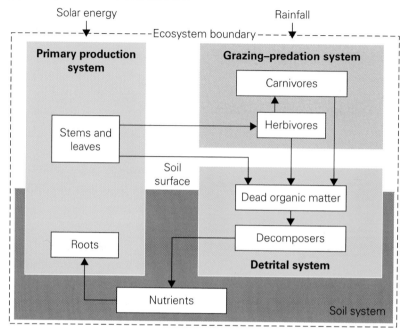

Ecosystems have three components:

- individual organisms, species and populations
- ecological processes, such as energy flow or ecological succession
- properties such as stability and fragility

Strictly, ecosystems have no boundaries because all parts of Earth are interconnected. However, a biogeographer usually has to define the limits of the ecosystem being studied, which can vary in size from a small garden pond to large areas such as the tundra of Alaska. Some studies regard the Earth as one ecosystem, variously called the **biosphere** or **ecosphere**. The ecosystem concept now occupies a central place in biogeography.

Energy flow in ecosystems

Photosynthesis

Energy flows through various levels in the biosphere after being captured by the process of **photosynthesis**:

$$6CO_2 + 6H_2O \xrightarrow{\text{light}} C_6H_{12}O_6 + 6O_2$$

carbon dioxide + water $\rightarrow$ glucose + oxygen

Photosynthesis converts solar energy into chemical energy. Solar radiation in the visible wavelengths, 0.4–0.7 μm (microns), provides the energy for carbon dioxide and water to be transformed into glucose and oxygen in **chloroplasts**, which are small structures within the cells of green plants. To take up carbon dioxide the plant needs to open pores (**stomata**) in its leaves. This leads to a water loss via the stomata by **transpiration**. Therefore, plants have to regulate stomatal opening in order to balance carbon dioxide uptake and water loss. This is particularly true of plants growing in dry areas. This is one reason why only 1–3% of solar energy hitting a leaf is used to produce carbohydrate.

Trophic structure

The term '**trophic**' (Greek, *trophe*, food) is used by biogeographers in their studies of the complexity of energy flows, or 'feeding relationships', in ecosystems — hence the terms **trophic level**, **trophic hierarchy**, **autotrophic** and **heterotrophic**. Organisms that transform light energy into chemical energy are **autotrophs** ('self-feeders') and include algae and some bacteria, as well as most plants.

- Photosynthesisers (autotrophs) are **primary producers.**
- Herbivores eat plants. They are **primary consumers**.
- Herbivores, carnivores, omnivores and decomposers are all **heterotrophs** ('general feeders'). They feed on the carbon already assimilated by autotrophs (Figure 1.2).

Figure 1.2 **Food chain at the edge of sea ice in the Arctic Ocean**

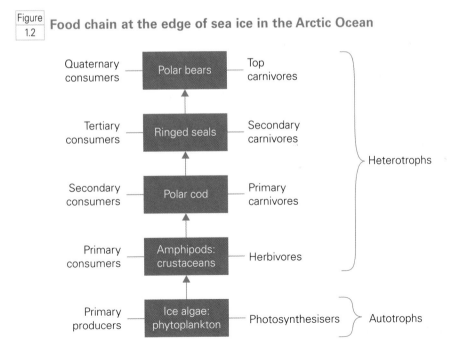

The term **food chain** is popular in the media. However, the idea of linear flows of energy in food chains is an oversimplification. In reality, energy flow within an ecosystem is complex, with many organisms occupying more than one trophic level. The trophic hierarchy is better viewed as a **food web**, rather than a chain. Figure 1.3 is a simplified representation of part of the food web for the northeastern Pacific Ocean off Alaska.

Figure 1.3 A marine food web in Alaskan waters of the northeastern Pacific

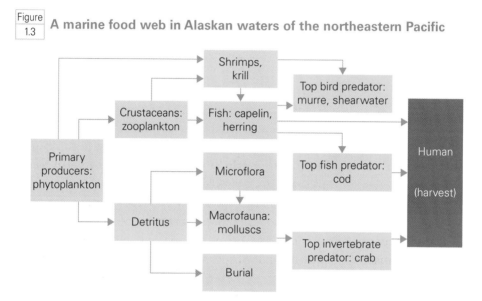

The trophic hierarchy is apparent, with planktonic plants at the base and humans at the top. Krill have the largest biomass in this food web and are the **keystone species** on which the entire ecosystem depends (Figure 1.4).

Figure 1.4 Krill: the key species in food webs of polar and subpolar regions

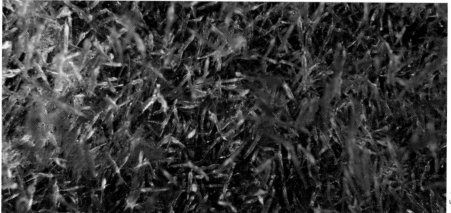

TopFoto

Eltonian pyramids

As a student in the 1920s, the Oxford biogeographer Sir Charles Elton visited the tundra ecosystem of Spitsbergen. He observed the feeding habits of insects, birds, and small and large mammals. He noted that predators are roughly ten times larger than their prey, but as the size of animals increases up the trophic hierarchy, their numbers decrease. Animals are larger and rarer further up the hierarchy. This is the **pyramid of numbers**.

Another of Elton's concepts was the **pyramid of biomass**, i.e. a pyramid representing the total mass of living material at each trophic level. The biomass of tundra birds is many times greater than that of their predators (e.g. the arctic fox) in the next highest trophic level (Figure 1.5). Elton was unsure why these pyramids exist. Is it explained by the size of the food or perhaps by the rate of reproduction?

Lindeman's energy studies

In the 1940s, the American biogeographer Raymond Lindeman was the first to recognise that it is the flow of *energy* within food webs that is the key-controlling factor. Using sampling techniques to measure the entire energy flow in a small lake in New England, he derived general laws to explain the structure and functions of ecosystems. His laws and equations are the basis of modern ecology.

He recognised that the amount of energy available to higher trophic levels declines because all organisms lose heat energy through **respiration** (Figure 1.5(c)). The overall equation for respiration is the reverse of that for photosynthesis:

$$C_6H_{12}O_6 + 6O_2 \rightarrow 6CO_2 + 6H_2O + \text{energy}$$
$$\text{glucose} + \text{oxygen} \rightarrow \text{carbon dioxide} + \text{water} + \text{heat}$$

Respiration occurs in the cells of all living organisms, all the time. It is not the same as breathing.

Figure 1.5 **Ecological pyramids of number, biomass and energy**

(a) Pyramids of number per 0.1 hectare

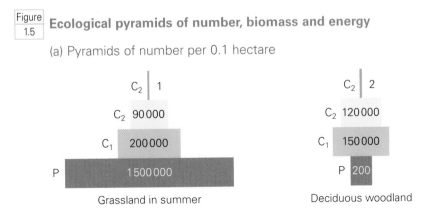

	Grassland in summer		Deciduous woodland
C_2	1	C_2	2
C_2	90 000	C_2	120 000
C_1	200 000	C_1	150 000
P	1 500 000	P	200

(b) Pyramids of biomass (dry mass, $g\,m^{-2}$)

Coral reef   English Channel

(c) Energy flow through the Silver Springs ecosystem, Florida ($kcal\,m^{-2}\,yr^{-1}$)

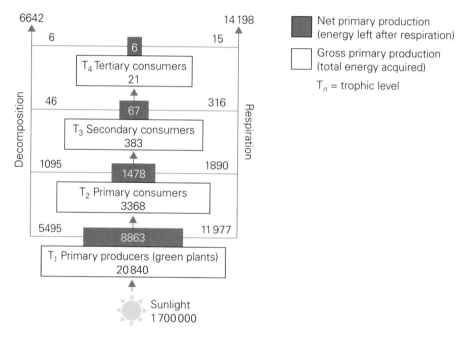

Activity 1

(a) Describe and explain the differences in the two pyramids of number shown in Figure 1.5(a).

(b) Describe and explain the differences in the two pyramids of biomass shown in Figure 1.5(b).

(c) Study Figure 1.5 (c), which shows energy flow through an aquatic ecosystem.

  (i) Describe the processes of energy conversion that occur at each trophic level in the ecosystem.

  (ii) Explain how and why the energy flowing from one trophic level (T_n) to the next highest (T_{n+1}) always decreases.

  (iii) Which trophic level is made up of herbivores, and which of carnivores?

Activity 1 (continued)

(d) Comment on the efficiency of energy conversion in the Silver Springs ecosystem, and describe its impact on animal populations at each trophic level.

The 10% rule

Roughly 10% of the energy received by one trophic level is passed on to the next. In other words, the higher level has 90% less energy for **metabolism** (all the chemical reactions in a living organism) than the lower level. This explains the pyramids of number and biomass, and *Why Big Fierce Animals are Rare*, to use the title of a popular Penguin paperback by Professor Paul Colinvaux. The precise shapes of the pyramids depend upon other factors, including the efficiency with which different organisms such as birds, mammals, insects and fish assimilate what they eat.

Two types of primary productivity

The rate of photosynthesis by plants is called **gross primary productivity (GPP)**. It depends upon light intensity, temperature, moisture and soil nutrients. Because of the energy loss in respiration, the amount of energy passed on to the next trophic level, the **net primary productivity (NPP)**, is equal to the gross primary productivity minus respiration, i.e. GPP – R = NPP. This is an important measure of how solar energy is made available to all organisms in ecosystems. It is expressed in units of the amount of carbon fixed per unit area per unit time — for example:

- grams per square metre per year ($g\,m^{-2}\,yr^{-1}$) (Table 1.1)
- tonnes per hectare per year (tonnes $ha^{-1}\,yr^{-1}$) (Table 5.1, p. 73)

Productivity in different biomes and ecosystems

Table 1.1 Typical values of net primary productivity

Vegetation region	Net primary productivity ($g\,m^{-2}\,yr^{-1}$)
Tropical rainforest	2200
Temperate deciduous forest	1200
Tropical savanna grassland	900
Boreal forest	800
Temperate prairie grassland	600
Tundra	600
Ocean	125
Desert	90

Activity 2

With reference to Table 1.1, explain the difference in NPP between a tropical rainforest and a temperate deciduous forest.

In land-based terrestrial ecosystems, NPP is highest in the humid tropics with their year-round strong radiation, high temperatures and abundant rainfall. Towards higher latitudes and in arid deserts, as these factors become less favourable, NPP decreases.

In the oceans, nutrients are the main limitation to the rate of photosynthesis by phytoplankton. The highest NPPs are found in temperate oceans where seasonal overturning of water brings nutrient-rich, deeper water to the surface or where nutrients are washed from the land into estuaries and onto continental shelves (Figure 1.6). Low productivity in marine ecosystems is offset to some extent by the more efficient transfer of energy from one trophic level to another.

Figure 1.6 The global distribution of net primary productivity

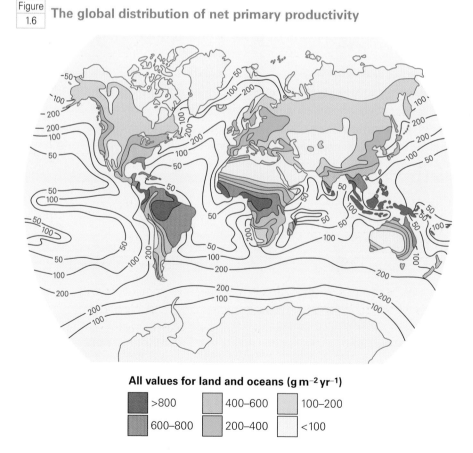

All values for land and oceans (g m^{-2} yr^{-1})

>800	400–600	100–200
600–800	200–400	<100

Activity 3

Table 1.2 shows data collected from 40 sites of natural vegetation over the Earth, ranging from tundra in polar latitudes to rainforest in the tropics. NPP ($g\,m^{-2}\,yr^{-1}$) was measured by cutting, drying and weighing the dry matter produced. Values for MAT (mean annual temperature, °C) were taken from nearby weather stations.

Table 1.2 Relationship between net primary productivity and temperature

Site	NPP ($g\,m^{-2}\,yr^{-1}$)	MAT (°C)	Site	NPP ($g\,m^{-2}\,yr^{-1}$)	MAT (°C)
1	290	−12	21	1030	8
2	280	−9	22	1050	14
3	490	−1	23	1080	14
4	530	0	24	1200	12
5	550	1	25	1300	9
6	550	3	26	1210	10
7	720	3	27	1450	10
8	780	−1	28	1290	10
9	820	5	29	1230	18
10	770	5	30	1610	4
11	1000	2	31	1510	8
12	990	3	32	1840	16
13	890	4	33	1510	14
14	875	4	34	2250	9
15	810	7	35	2520	14
16	815	8	36	2460	15
17	850	8	37	2810	27
18	1070	5	38	2820	28
19	1000	9	39	2130	22
20	1020	8	40	2520	25

(a) Plot the data as a scattergraph with MAT on the x-axis and NPP on the y-axis.

(b) Is the trend a line or a curve? Suggest reasons why it is this shape.

(c) Use Microsoft Excel to calculate the correlation coefficient between the two sets of data.

(d) Use Microsoft Excel to calculate, and then draw on the graph, the regression line between the two sets of data.

(e) State the regression equation. What is the average change in NPP for each rise of 1°C in MAT?

The cycling of nutrients

Plants and animals require **nutrients**. Green plants need 16 essential nutrients. These are divided into **macronutrients**, which are needed in relatively large quantities, and **micronutrients** or **trace elements**, which are needed in very small amounts. The macronutrients are carbon, hydrogen, oxygen, nitrogen, phosphorus, potassium, calcium, magnesium and sulphur. The micronutrients are iron, manganese, copper, zinc, molybdenum, boron and chlorine (Table 1.3). Carbon, hydrogen and oxygen are obtained from the atmosphere and water and are not discussed here.

Table 1.3 Plant nutrients

Macronutrient	Chemical symbol	Micronutrient	Chemical symbol
Nitrogen	N	Iron	Fe
Phosphorus	P	Manganese	Mn
Potassium	K	Copper	Cu
Calcium	Ca	Zinc	Zn
Magnesium	Mg	Molybdenum	Mo
Sulphur	S	Boron	B
		Chlorine	Cl

Plants absorb most nutrients as **ions**. These are atoms (or groups of atoms) with electrical charges. These ions are found in soils and occur as **cations** (ions with a positive charge) and **anions** (ions with a negative charge). They are absorbed by plant roots. Table 1.4 shows the ions of the six macronutrients.

Table 1.4 Nutrients as ions

Nutrient	Cation	Anion
Nitrogen	Ammonium (NH_4^+)	Nitrate (NO_3^-)
Phosphorus	–	Phosphate (PO_4^{3-})
Potassium	K^+	–
Calcium	Ca^{2+}	–
Magnesium	Mg^{2+}	–
Sulphur	–	Sulphate (SO_4^{2-})

Nitrogen (N), phosphorus (P) and potassium (K) are the most important of these nutrients for plants, and the ones most commonly added by farmers as fertilisers.

Cycles, stores and flows

'Cycling' means 'using over again'. Energy flows from one trophic level to another and there is substantial loss with each flow. Energy can never be returned to its source (p. 13). In short, energy *flows* whereas nutrients *cycle*. Theoretically, a nutrient ion could be moved around the biosphere indefinitely.

There are two parts to every nutrient cycle:

- **stores** of the nutrient (sometimes called **pools**)
- **flows** of the nutrient between the stores (Figure 1.7)

Figure 1.7 **Stores and flows in a generalised ecosystem**

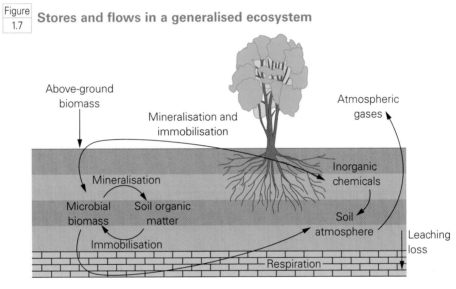

Nutrient cycles are different for each nutrient element. However, two broad classes are recognised:

- In **gaseous cycles**, for example those of nitrogen, carbon and sulphur, the main stores are gases under normal atmospheric conditions. Their main input into the ecosystem is through some method of biological fixation.
- In **sedimentary cycles**, for example those of potassium, phosphorus, calcium and magnesium, nutrients exist as solids or in solution and inputs originate from rock weathering.

Nitrogen gaseous cycle

Nitrogen (N) is an important plant nutrient. It is a component of DNA, amino acids and, therefore, proteins. Almost all (98%) of the nitrogen in soils is in the organic matter. Plant litter and humus are continually being broken down by soil microorganisms in a process called **mineralisation**. This releases nutrients, including nitrogen.

Organic nitrogen is mineralised to ammonium (NH_4^+) ions by soil microbes. The ammonium ions are then converted to nitrate ions — the ion absorbed by plants — in a two-stage process called **nitrification**. Each stage is performed by specific **chemoautotrophic** bacteria:

- **Stage 1** Ammonium is converted to nitrite by *Nitrosomonas*:
  $$NH_4^+ \rightarrow NO_2^-$$
- **Stage 2** Nitrite is converted to nitrate by *Nitrobacter*:
  $$NO_2^- \rightarrow NO_3^-$$

Nitrification reactions are oxidation reactions.

In the absence of oxygen in waterlogged soils and marshes, ammonium, nitrite and nitrate ions are reduced to nitrogen gas (N_2). This process is called **denitrification**. It returns nitrogen to the atmosphere and is a major loss of nitrogen to plants.

Most organisms cannot use atmospheric nitrogen. However, some micro-organisms (**nitrogen-fixers**) can. Some fixation is carried out by free-living soil microbes but most is by bacteria that form a symbiotic relationship with plants called **legumes**. Bacteria of the genus *Rhizobium* occupy nodules in the roots of leguminous plants, from which they obtain carbon. In return, they fix atmospheric nitrogen into a form that the plant can use. This is one of the main flows by which atmospheric nitrogen becomes usable (Figure 1.8).

| Figure 1.8 | **The nitrogen cycle** |

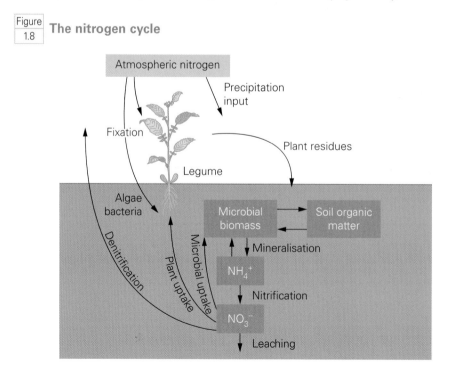

'**Green manuring**' is the farming technique of growing legumes for the nitrogen they can add to the soil. Increased additions of nitrogen to soils as a result of pollution are covered later (p. 64).

Sedimentary cycles
Cation exchange

In order to understand sedimentary nutrient cycles and soil chemistry, it is necessary to understand the process of **cation exchange**, of which it has been said: 'after photosynthesis, cation exchange is the second most important reaction in agriculture' (see p. 39). Clay and humus colloids in soils have enormous surface areas that carry **negative charges**. **Positively charged cations** are attracted to the electrostatically negative colloid surfaces. The cation 'cloud' has two parts. The first is a tightly held layer at the colloid surface, called the **Stern layer**; the second is the **diffuse layer** (Figure 1.9).

Figure 1.9	**The double-layer chemistry of clay minerals**

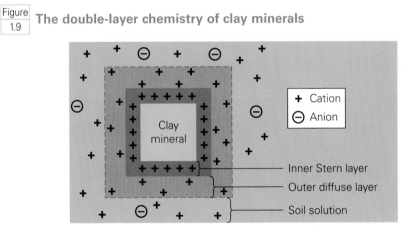

The process of colloids holding ions in this way is called **adsorption**. Cations in the diffuse layer can exchange with those in the inner layer by cation exchange (Figure 1.10).

Figure 1.10	**Diagrammatic representation of cation exchange**

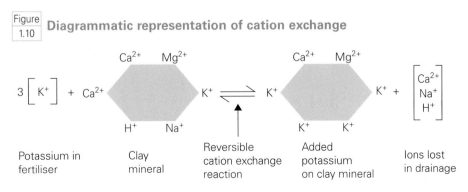

Plant absorption

There is a state of dynamic equilibrium or balance between:

- nutrients adsorbed on the colloids
- nutrients in the soil solution
- nutrients absorbed by roots

Cations held in the Stern layer move into the soil solution if the concentration in the soil solution is lowered — for example, through removal by plants or by leaching. The adsorbed cations are thus a store or 'bank balance' of nutrients, supplying the plant when amounts in the soil solution run low (Figure 1.11).

Figure 1.11 **Processes of nutrient uptake in plant–soil systems**

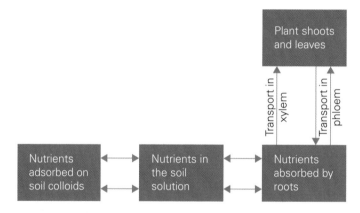

Potassium sedimentary cycle

Potassium (K) is important for cell sap, enzymes, photosynthesis and disease resistance in plants. It occurs in soils, and is absorbed by plants as the cation K^+. In soils, it is present as:

- a store of non-available potassium ions in rock minerals, such as feldspar and mica
- exchangeable potassium ions on clay and humus colloids
- potassium ions in soil water

Figure 1.12 shows the nutrient cycle of potassium in the New Guinea rainforest. The store of potassium ions in rock minerals is released by weathering and enters the soil-solution store. However, it is leached quickly away and plants obtain little potassium from this store. The main input into the soil is from the vegetation. Potassium ions in plants are recycled by leaf-fall and litter-fall and subsequent mineralisation by microorganisms. A second significant input comes from the **canopy leaching** of potassium from living leaves, which makes the concentration of potassium ions in throughfall greater than that in rainfall. There is a small atmospheric input in rain and particulates.

Figure 1.12 **Potassium (K) nutrient cycle in the New Guinea rainforest**

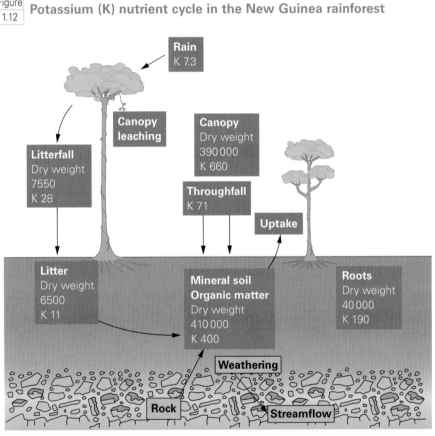

Dry weight values: kg ha^{-1}
K flows: kg ha^{-1} yr^{-1}

Phosphorus sedimentary cycle

Phosphorus (P) is important for photosynthesis, nitrogen fixation and root development in plants; it is a component of DNA. It is present in both soil organic matter and soil minerals. The relative proportions depend upon the organic content. In organic peat soils, 80% of phosphorus is in the organic matter and is released by decomposition in mineralisation. In non-organic soils, most phosphorus is in soil minerals and is released by weathering. The sedimentary cycle of phosphorus differs from most others because mineralisation and weathering produce phosphate (PO_4^{3-}) ions, so the cycle is based on an anion, rather than a cation.

Phosphate ions form relatively insoluble chemical compounds in soil. This reduces phosphate's availability to plants. In acid soils, insoluble iron and aluminium phosphates are formed; in alkaline soils, insoluble calcium phosphate is formed. These reactions are called **phosphorus fixation**. They limit plant productivity because phosphate is taken out of the active cycle. The content of available phosphorus as phosphate ions in most soils is small, as is the leaching loss. Farmers add fertilisers that contain phosphorus. However, these might be ineffective unless the soil pH is near to neutral (pH 7), i.e. where the content of iron, aluminium and calcium is lowest. Soil pH is discussed in Chapter 3.

Importance of soil organisms in nutrient cycling

Soil microorganisms play several important roles in ecosystems:

- Specialised groups of chemoautotrophs are needed for active nitrogen and sulphur cycles.
- Plant and animal residues added to the soil surface are broken down to release the nutrients they contain. Decomposer organisms quickly colonise residues and release ions, through mineralisation, so that they are recycled and used by plants.
- Microorganisms themselves need nutrients. Therefore, if a nutrient is in short supply in the soil, plants can suffer through competition from the soil microbes.
- Dead microorganisms are an important part of soil humus.

It is difficult to overemphasise the importance of soil organic matter and microorganisms in ecosystems. Ecosystems, like soils, are not static and inert. They are dynamic and living, completely dependent on active nutrient cycles. Soils would soon become infertile if nutrients were not continuously recycled by microbes.

2 Geographical distribution of plants and animals

One aim of biogeography is to explain the distribution of plants and animals on the Earth. Why do polar bears live only in the cold Arctic? Why are palm trees associated with tropical latitudes?

The portion of the Earth occupied by a species of plant or animal is the **range** of that species. The factors that define ranges are:

- physical environment
- chemical environment
- biological influences
- human actions
- history

The role of humans is clearly biological, but its importance makes it worth analysing separately (pp. 33–34). The 'history' factor is discussed in Chapter 7; the remaining factors are discussed in this chapter. In the real world, the distribution of organisms is controlled not by a single factor but by a complex combination of several factors operating together (Figure 2.1).

| Figure 2.1 | Factors controlling species distribution |

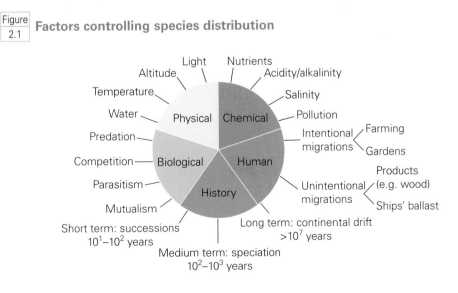

Physical and chemical environment

Light

Plants need light for photosynthesis. They are **phototropic**, i.e. they grow towards the light. However, there are exceptions. For example, some woody lianas in tropical rainforests (e.g. the Swiss cheese plant) are **skototropic**, i.e. they seek out the darkness of a tree's shade to climb.

Heliophytes grow best in full sunlight; **sciophytes** grow best in shade. For example, pines, birch and many species of herbs and grass are shade intolerant, whereas lime, elm and beech are shade tolerant. Shade-intolerant species are often pioneer plants in vegetative successions and colonise rapidly any gaps that occur in forests. Shade-tolerant species are able to germinate below a tree canopy. They are usually long-lived perennials, and form climax species in successions. Succession is covered in Chapter 7.

Temperature

Temperature is a major control factor of rates of photosynthesis, metabolic activity and growth. **Chilling stress** and **freezing stress** occur when temperatures fall below the minimum needed by a particular species. Low winter temperatures are important in restricting the poleward limits of many plants. This is true not only for tropical plants, whose flowers are killed by freezing, but also for many temperate zone plants, such as the coniferous coastal redwoods of California, which die below −25°C. Low summer temperatures explain the position of the circumpolar tree line that correlates with the 10°C July isotherm (p. 96).

Plants have adaptations to survive hostile temperatures. For example:
- deciduous trees lose their frost-sensitive leaves in autumn
- evergreen conifers survive winter temperatures as low as −50°C by making sugars in their cells that act as a kind of antifreeze
- in deserts, **dimorphic** plants, such as mesquite, survive excessive summer heat by growing small leaves in the hot season to replace the larger leaves of the cool season

Water

Water is essential for all biochemical processes in organisms. In plants it is important in maintaining rigidity and in cooling leaves by transpiration. Plants can be classified by their need for water:
- **Xerophytes** survive in arid areas.
- **Mesophytes** require moist conditions.
- **Hydrophytes** grow in water or very wet soils.

Xerophytes survive in deserts and in seasonally dry conditions using the strategies shown in Table 2.1. A baobab tree, which stores water, is shown in Figure 2.2.

Table 2.1 Survival strategies of xerophytes

Strategy	Adaptations	Examples
Drought-escapers	Deciduous	Annual grasses and herbs
	Above-ground die-back	Sedges
Drought-avoiders	Different seasonal leaves	Ocotillo (southwest USA)
	Waxy leaves (**sclerophyllous**)	Oleander (Mediterranean)
	Deep roots	Mesquite (southwest USA)
	Water storage in tissue	Baobab tree (Africa)
	No leaves, so photosynthesis through stems	Cacti (New World), many shrubs e.g. retama (Mediterranean)
	CAM photosynthesis	Cacti (New World)
Drought-tolerators	C_4 metabolism plants	Pea family, grasses
	Resurrection plants	*Welwitschia* (Namibia), mosses

Figure 2.2 **The spongy wood of the baobab tree allows water storage**

Paul L. H. Cook (http://langabi.name)

Acidity

Plants are sensitive to the chemical conditions of soil and water, though these are important at the local, rather than global, scale. **Calcifuges** (lime-hating plants) such as ling and rhododendron grow only on acid soils because the calcium present in alkaline soils prevents the plant from absorbing iron. In contrast, **calcicoles** (lime-loving plants) need plenty of calcium in the soil, as shown by alpine plants growing on base-rich rocks in Wales, the Pennines, the Lake District and Scotland (Figure 2.3).

| Figure 2.3 | Mountain avens is an arctic–alpine calcicole, here growing on Durness limestone, Assynt, Scotland |

Michael Raw

Some plant groups, for example lavenders and bedstraws, have some species that are calcifuges and some that are calcicoles. Thus, these species are good indicators of the degree of soil acidity. In general, tropical plants are more tolerant of acidity than those that have evolved on the higher pH soils of the mid-latitudes.

Salinity and other chemical factors

Salinity is lethal to most plants because water flows out from plant cells into saline soil water. **Halophytes** are plants that can tolerate salty soils in salt marshes and deserts (Figure 2.4).

Figure
2.4 **Halophytic plants growing in a saline soil, Turkey**

Ken Atkinson

Plants are also sensitive to other chemicals — for example, the high concentration of magnesium in soils on dolomite limestone and of nickel and other heavy metals in soils on serpentine. Vegetation is sparse on soils contaminated by industry that contain heavy metals, such as nickel, copper, lead and mercury. It is restricted to tolerant 'metalliferous' plants.

Biological influences

Plants and animals interact with other species and these interactions influence their geographical ranges. Three types of interaction are shown in Table 2.2.

Table
2.2   Types of interaction between organisms

Interaction	Subtype	Nature
Predation	Herbivore	Stenophagous predators
	Carnivore	Lotka–Volterra theory (p. 28)
Competition	Interspecific	Competition between two or more species
	Intraspecific	Competition within a single species
Symbiosis	Mutualism	Benefit to both species
	Parasitism	One species benefits and the other suffers

Predation

Predators that have a single prey, or a narrow range of prey, are called **stenophagous**. This is common in plant-eating insects and larvae — for example, the caterpillars of butterflies. There are also examples in mammals, such as the dependence of the giant panda on bamboo. Stenophagy might appear to be a poor strategy because the distribution of the predator is limited by the distribution of the prey. However, as plants often have toxins to deter herbivores, the predator will have evolved resistance to these toxins and can feed on the plant with little competition from other herbivores. In Britain, the dependence of the caterpillars of the cinnabar moth on ragwort is an example.

If predation becomes heavy, the prey population will 'crash'. This, in turn, reduces the predator population. The prey population then recovers, leading to an increase in the predator population. This is the **Lotka–Volterra** theory, which explains linked oscillations in prey–predator populations. In the ecology of the Arctic, the cyclical nature of interactions between snowshoe hare and lynx, between willow twigs and snowshoe hare, and between lemming and snowy owl have been studied in depth. Sir Charles Elton's study of the 1845–1935 records of the Hudson's Bay Company, Canada, show clearly the cycles of prey (hare) and predator (lynx) (Figure 2.5).

Figure 2.5 | **Lotka–Volterra cycles in snowshoe hare and lynx populations in the Canadian Arctic**

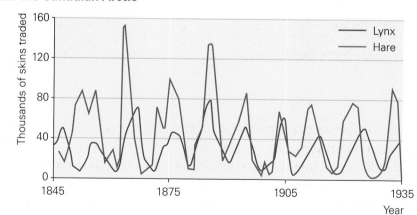

Competition

Charles Darwin placed great emphasis on **competition** in limiting the distribution of species. The first type of competition is that for resources, such as light, food, water and nesting sites. A second type is called **allelopathy**. This is the exclusion of one species by another by means of toxins, usually terpenes, exuded from the roots. There are many suspected cases of allelopathy on Earth, in particular where the regular spacing of plants cannot be explained by root competition or shading (Figure 2.6). Those that have been well studied are the 'halos' of bare ground around creosote bushes in the deserts of the American southwest and around sagebrushes in both Old World and New World deserts. In Britain, the bracken fern releases toxic chemicals and spreads at the expense of competitors such as heather.

Figure 2.6 **Spacing of the bunch grass, esparto, caused by allelopathy, southern Spain**

Ken Atkinson

Symbiosis

Associations between two species can influence the range of both. **Symbiosis** is an interaction between species (plant–plant, plant–animal, or animal–animal) that has developed over time by **coevolution**. The symbiosis is usually necessary for one of the species to survive. Interactions that benefit both species are **mutualistic**; those in which one benefits at the expense of another are **parasitic**.

Mutualism

Mutualistic associations between plants and insects are common. In southwest USA, moths of the genus *Tegeticula* feed only on the yucca plant and, thus, pollinate them. Neither species can extend its range into areas that cannot support the other. The association between *Rhizobium* bacteria and legumes, which leads to the fixation of atmospheric nitrogen (p. 18), is one of the most important in ecology and agriculture, though in this association each species can exist independently of the other.

Parasitism

The plants *Cynomorium coccineum* and *Cytinus hypocistis* and some plants of the genus *Cuscuta* are parasitic on Mediterranean plants (Figure 2.7). *Rafflesia arnoldi*, with the world's largest flower at almost 1 metre in diameter, is parasitic on woody vines in some tropical forests. Most parasites contain no chlorophyll; mistletoe in temperate woodlands is a semi-parasite as it makes its own carbohydrate by photosynthesis, while taking water and nutrients from the host tree. In animals, variations in the abundance of parasites and hosts may show Lotka–Volterra cycles. For example, red grouse in Britain suffer high mortality rates from a nematode worm on a 5-year cycle.

| Figure 2.7 | Cynomorium: a chlorophyll-free Mediterranean parasitic plant |

©blickwinkel/Alamy

Ecological tolerance and ecological niches

Every living species has evolved within the limits set by the natural environment, which has two components:

- The **abiotic component** is that part of the environment that is not biological, i.e. climate, geology, water and soil (though in some senses, the soil is biological).
- The **biotic** component is the biological environment, i.e. living organisms.

The noun 'ecology' and the adjective 'ecological' always cover both the abiotic and the biotic components.

Species tolerance curves

Shelford's law states that for each controlling abiotic factor affecting the range, organisms have a minimum and a maximum limit of tolerance. A **species tolerance curve** (Figure. 2.8) can be drawn for each controlling factor (e.g. soil acidity, mean temperature or rainfall). It is usually bell-shaped, with the peak representing optimal growing conditions. The *x*-axis of the species tolerance curve can be thought of as an **environmental gradient**.

Figure 2.8 **A species tolerance curve**

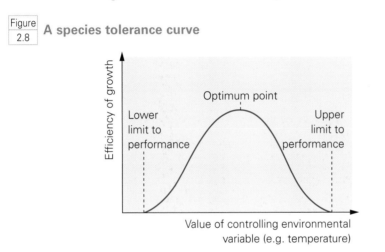

In the real world, species can compete with one another. Figure 2.9 shows the species tolerance curves of two communities of vegetation. In one community there is strong interspecific competition; in the other, interspecific competition is weak and the different species can coexist.

Figure 2.9 **Strong and weak competition between species**

(a) Strong competition

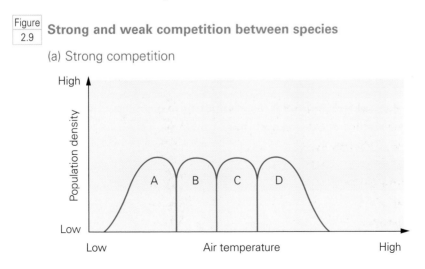

(b) Weak competition

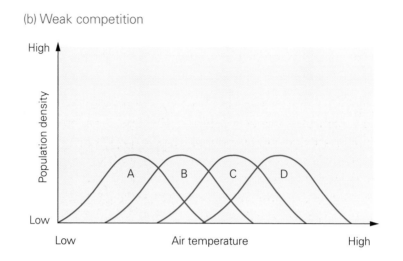

Ecological niches

Every organism in an ecosystem has its own **niche**. A niche includes the organism's function as well as its location. In 1835, Charles Darwin observed variation in beak size and shape among different species of finch on the Galapagos Islands. However, it was over a century later before it was discovered that the different species of finch have different beak sizes for eating different sizes of seed. Competition forces each finch species to specialise on one seed source and this specialisation allows coexistence.

The coexistence of so many mammals in the tropical rainforest is also partly due to specialised niches in both time and space. The tall trees provide a variety of living spaces and foods and some species only live in a single layer in the forest (see Chapter 5). Some animals are active only by day; others only by night.

While the niche concept explains coexistence in animal communities, it is more difficult to apply to plants, which have broadly similar growth require-ments. Although there are differences in such things as size, canopy and rooting depth, it is not clear how hundreds of different tree species in tropical forests reflect specialised niches. This and other questions related to biodiversity are discussed in Chapters 5 and 8.

Activity 1

Describe the ecological niches that are found in a deciduous woodland in lowland England. What plants and animals occupy these niches?

The influence of human activities

Aliens and introduced species

In biogeography, an **alien** is a species that has reached a particular area as a consequence of human activities. An **introduced species** is one whose range has been deliberately expanded by human actions. There are few parts of the Earth that are not affected by alien species, whether intentionally or unintentionally. European contacts and settlement in the 'colonial lands' of the western hemisphere, southern Africa and Australasia over the past few hundred years have resulted in the high proportion of aliens in these countries. The prickly pear in the Mediterranean is one of a number of species of plants brought back from Latin America by Spanish conquistadors. However, only about 5% of plant species in Europe come from elsewhere (Figure 2.10). In contrast, 30% of plants in North America are aliens.

Figure 2.10 **Agave, or century plant, in Spain: an import from Mexico**

©Ellen McKnight/Alamy

Alien invaders are usually successful in competition with native species, according to the **principle of competitive displacement**. In the absence of natural competitors and enemies such as insects, fungi and diseases, aliens have much larger niches than they occupy in their native habitats. They have been aptly called 'unfettered demons'. In Florida, there are 4000 plant species, of which 1200 are aliens; 90% were deliberately introduced as crops, aquarium plants, windbreaks and garden ornamentals. After Hawaii, Florida has the worst problems with aliens in the USA; 62 have been listed as 'unwanted weeds' by the Florida Exotic Pest Plant Council. These include kudzu vine, lantana, water hyacinth, Australian pine, paperbark, Brazilian pepper, laurel fig and guava.

Controlling alien species

Biological control by natural enemies imported from the alien's native range can be a successful way of managing problem aliens. Alligator weed in Florida has been reduced to 1% of its former extent by the introduction of three insect grazers from its indigenous home in Argentina. Before any such introduction, experiments have to be carried out to ascertain that the predator will not attack native species in preference to the alien target.

Accidental introductions

The River Tweed in Scotland provides a good example of non-deliberate introduction. In the nineteenth and twentieth centuries, wool was imported from Australasia, South Africa, Russia, Asia and South America for the local woollen industry. Seeds in the wool were washed into local rivers. A plant survey in 1919 listed 348 'wool aliens' on the banks of the Tweed. All that now survives is one New Zealand rose on nearby Holy Island, Northumberland. Why has the principle of competitive exclusion not worked here? Why, given the thousands of exotic plants grown in European gardens, are there not more 'garden escapes' in Europe's wild vegetation?

There appear to be two reasons. The first is climate; the cooler the climate, the harder it is for aliens to survive in the face of competition. The second lies in habitat change brought about by the farming systems that were introduced into colonial countries. European settlers raised European animals and used European grasses and other agricultural plants. In this way, they prepared the ground for invasions of alien plants from their homelands.

Activity 2

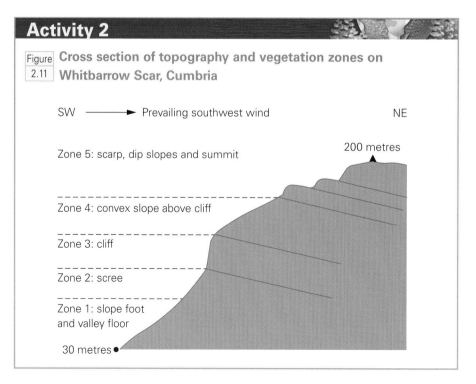

Figure 2.11 Cross section of topography and vegetation zones on Whitbarrow Scar, Cumbria

SW ⟶ Prevailing southwest wind   NE

Zone 5: scarp, dip slopes and summit   200 metres

Zone 4: convex slope above cliff

Zone 3: cliff

Zone 2: scree

Zone 1: slope foot and valley floor

30 metres

Activity 2 (continued)

Study Figure 2.11, which shows vegetation zones on the slopes of Whitbarrow, Cumbria.

(a) Describe and explain the *physical factors* operating in each zone that will influence the type and distribution of the vegetation.

(b) Using Figure 2.1, discuss what *non-physical* factors might also be important on Whitbarrow Scar.

Activity 3

Discuss, with examples, how humans have affected the distribution of plants and animals across the globe.

3 Soils: their properties and formation

Soils are an important part of the biosphere. They provide nutrients, water and physical support for crops and other plant communities. In return, plants provide organic matter as litter to the soil surface, which is necessary for a living soil community, in order to improve soil physical properties and to recycle nutrients.

Physical, chemical and biological properties of soils

Soils consist of inorganic mineral material, organic matter, water and gases. About 45% of the volume of an 'average' soil is made up of minerals, 5% is organic matter and the remaining 50% is pore space. The pores are filled with air or water; the exact percentages are dependent on recent precipitation. Under severe drought conditions, the water content might be as low as 5%; in saturated soils, it might be 35% (Figure 3.1).

 Volume composition of a typical soil

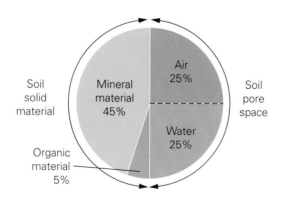

Air and water components are variable (shown by the dashed line)

Texture

The inorganic minerals in soil are of two types:
- **Primary minerals** are the residues from the weathering of the underlying rock. They consist mainly of resistant minerals such as quartz and feldspar and comprise large, sand-sized particles.
- **Secondary minerals** are clays and oxides of iron and aluminium that have been newly formed within the soil by chemical reactions between weathering products. Because of their small size (<0.002 mm diameter), secondary minerals are called **colloidal minerals**.

The proportions of mineral particles in different size-ranges is the **soil texture**. This physical property is detected by sieving, or simply by feeling moistened soil between the fingers. In the UK classification:
- **stones and gravel** particles are larger than 2 mm diameter
- **coarse sand** particles are 2.00–0.2 mm
- **fine sand** particles are 0.2–0.06 mm
- **silt particles** are 0.06–0.002 mm
- **clay** particles are <0.002 mm (<2 microns)

A balanced mixture of sand, silt and clay is a **loam**.

Once the percentages of the different size-ranges have been determined, the soil textural triangle allows the texture to be classified (Figure. 3.2).

| Figure 3.2 | **The soil textural triangle** |

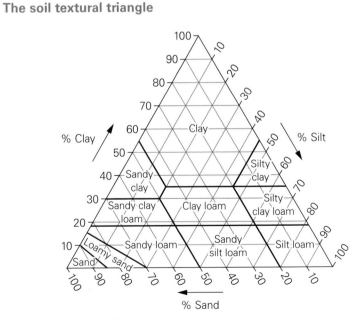

Texture is important because of its influence on the water-holding and aeration properties of soils. The larger the particle size, the larger are the gaps (**soil pores**) between particles. The total pore space — **soil porosity** — is greater in clay soils than in sandy soils. However, the large pores in sands increase the rate of drainage of water and decrease the amount of water held in soil-storage for plants (the **water-holding capacity**). Hence, sands drain rapidly and are prone to drought; clays drain slowly and are prone to waterlogging.

Activity 1

Particles	Soil A	Soil B	Soil C	Soil D	Soil E
Sand (%)	30	18	25	80	55
Silt (%)	40	54	20	15	20
Clay (%)	30	28	55	5	25
Total (%)	100	100	100	100	100

(a) Using the soil textural triangle (Figure 3.2), place each of soils A, B, C, D and E into a texture class.

(b) Rank the soils, from highest to lowest, according to the surface area of the mineral particles that is available for chemical reactions such as cation exchange.

(c) Rank the soils, from highest to lowest, according to their water-holding capacity.

Colloidal properties

Soil colloids are mineral particles (clays) that are less than 2 microns in diameter. They are that part of the organic matter referred to as **humus** or **humic colloids**. Colloids have two properties that are important for soil fertility. First, they possess a large surface area. Consider a cube of edge 1 metre. It has a surface area of $6\,m^2$. If this cube is subdivided into smaller cubes of edge length 1 millionth of a metre, the total surface area becomes $6\,000\,000\ m^2$. The finer the soil texture, the greater is the surface area.

Second, the surfaces of colloids are electrostatic, i.e. they carry electrical charges. The charges arise from the chemical composition of the ions that make up the colloids (the **permanent charge**) and the activation of ions on colloidal surfaces by the surrounding medium (the **variable charge**). The latter depends mostly on soil pH. Although negative (–) and positive (+) charges both occur on colloids, the negative outweigh the positive, giving a net negative charge.

Cation exchange

Colloids, like all mineral particles in soils, are surrounded by a film of water, the **soil solution**, which contains chemicals as dissolved ions. The negatively charged colloids attract positively charged ions (cations) from the soil solution to their surface. This produces negatively charged colloids of clay and humus, each with a tightly held layer of positively charged cations. The dominant soil cations are calcium (Ca^{2+}), magnesium (Mg^{2+}), potassium (K^+), sodium (Na^+) and hydrogen (H^+). They are called 'base cations'. Calcium, magnesium and potassium are important plant nutrients.

Cation exchange is an important soil process whereby cations held at the colloid surface exchange with cations in the soil solution (Figure 1.10, p. 19). If a farmer adds a potassium fertiliser to the soil, the level of potassium ions in the soil solution is increased and potassium ions will exchange with cations present on the colloidal surfaces. The important point is that the potassium ions will be held tightly and not subject to loss by leaching. However, they are available to plant roots. Therefore, colloids provide a store for nutrients until they are required by plants. The ability of soils to store nutrients in this way is called the **cation exchange capacity** (CEC). Soils with large clay and humus contents have high CECs; soils with sandy textures and low humus contents have low CECs.

Soil pH

Acidity is an important soil property. Acids have high concentrations of hydrogen ions (H^+) — the more hydrogen ions, the stronger the acid. The **pH scale** is a logarithmic scale of the hydrogen ion concentration, defined as: 'the negative index of the H^+ concentration'. The scale runs from 0 to 14; the neutral value is pH 7. Values below 7 denote acidity. The lower the value, the greater the hydrogen ion concentration and acidity. A pH value above 7 denotes alkalinity. Most soils are in the pH range 5.5–7.5, although there are exceptions — for example, podzols (approximately pH 3.0) and alkaline soils (around pH 9).

Over time, under natural conditions, soils become more acidic because leaching by rainwater of pH 5.5 removes calcium, magnesium, potassium and sodium ions and increases the proportion of hydrogen ions. Hydrogen ions are also produced when organic matter is decomposed in **humification**. Some farming practices increase acidity by removing base cations in crops and by adding fertilisers. Liming materials are added to raise pH levels.

The effects of pH on soil fertility

High acidity is detrimental to:

- the activity of microorganisms
- the availability of nutrients for plants
- soil structure formation, which depends on calcium ions

Extreme acidity values (pH <2) indicate the presence of metal pollutants. It is now known that aluminium dissolves from rock minerals at pH 4 and aluminium ions enter the soil solution. This causes soil fertility problems because:

- aluminium ions occupy exchange sites on colloids and therefore reduce the base cations
- the hydrogen ion concentration in the soil solution is increased

Aluminium ions are also toxic to plants and animals.

Soil structure

Sand, silt and clay rarely exist as individual particles in a **single-grained** state. Usually they coalesce into **soil aggregates**; this state of aggregation is called the **soil structure** (Figure 3.3).

| Figure 3.3 | **Types of soil structure** |

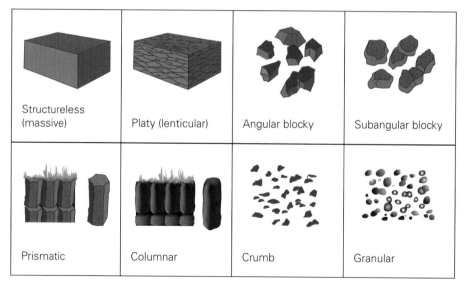

| Structureless (massive) | Platy (lenticular) | Angular blocky | Subangular blocky |
| Prismatic | Columnar | Crumb | Granular |

Physical, chemical and biological processes are responsible for aggregation (Table 3.1).

Table
3.1 Processes involved in forming the structure of soils

Type of process	Process
Physical	Frost action
	Wetting and drying
	Build up of pressure from growing roots
	Timely ploughing and harrowing
Chemical	Bonding of clays to the cations Ca^{2+}, Mg^{2+} and Al^{3+}
	Cementing by iron oxide, Fe_2O_3
Biological	Production of 'gums' of sugars by microorganisms
	Binding by fine roots
	Binding by hyphae of fungi

Good soil structure is the key to a fertile soil. Such a structure is one that:

- is not too compacted
- has large pores for infiltration and drainage of water, and for aeration
- has small pores for holding water in the soil
- is stable

In a stable soil, the colloids stick together. This is called **flocculation**. When clay colloids exist as individuals, they are in a **dispersed** state. Dispersion occurs when aggregates fall apart in water or under mild pressure.

A poor soil structure is one in which the aggregates break down under rainfall-impact or cultivation to give either compacted surface crusts or single grains that are removed easily by wind, as in the 'dust bowl' that occurred in the USA and Canada in the 1930s.

The role of inorganic and organic colloids in forming and maintaining a good structure is significant. Individual clay particles are plate-like in shape and cations on their surfaces, particularly those of calcium, magnesium and iron, act as bridges and enable the clay plates to stick together into **clay domains**.

However, it is biological activity in soils that produces the most stable structures (Table 3.1). Agricultural practices that do not provide a continuous supply of organic matter are likely to produce unstable soils. This is covered in Chapter 4.

Bulk density

Bulk density is a good indicator of soil structure. This is the **mass of soil per unit volume**, usually measured in $g\,cm^{-3}$. It is measured by taking a known volume of soil, drying it and weighing it. The lower the bulk density, the greater is the soil porosity. Dense, compacted soils have values $>1.5\,g\,cm^{-3}$; values less than $1.0\,g\,cm^{-3}$ indicate porous soil structures.

Factors of soil formation and the zonal soil concept

In the nineteenth century, the Russian soil scientist Dokuchaev was the first to explain why soils vary in different places on Earth. He realised that soil variation is produced by the interaction of four **factors of soil formation**:

- climate
- vegetation
- parent material
- topography

'Organisms, including human activity' is a fifth soil-forming factor and time is sometimes listed as a sixth. The effects of the soil-forming factors on soil properties are illustrated in Figure 3.4.

| Figure 3.4 | **The effects of the factors of soil formation on soil properties** |

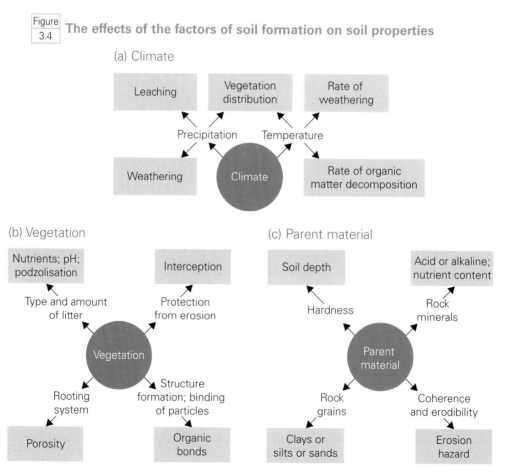

(a) Climate

(b) Vegetation

(c) Parent material

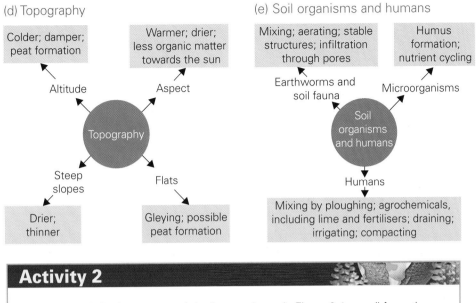

(d) Topography

Colder; damper; peat formation

Warmer; drier; less organic matter towards the sun

Altitude

Aspect

Topography

Steep slopes

Flats

Drier; thinner

Gleying; possible peat formation

(e) Soil organisms and humans

Mixing; aerating; stable structures; infiltration through pores

Humus formation; nutrient cycling

Earthworms and soil fauna

Microorganisms

Soil organisms and humans

Humans

Mixing by ploughing; agrochemicals, including lime and fertilisers; draining; irrigating; compacting

Activity 2

Discuss the relative importance of the factors shown in Figure 3.4 on soil formation.

In reality, the factors interact and do not act independently of each other. Thus, an acidic parent material such as sandstone favours an acid-tolerant vegetation such as heather, the litter from which further lowers the soil pH (Figure 3.5).

Soil zonality

Dokuchaev classified soils into three groups:

- **Zonal soils** are well-developed soils that are associated with climate and vegetation regions (**natural regions**, the equivalent of **biomes** — Chapter 5).
- **Intrazonal soils** are soils in which one local factor dominates. The two main local factors are:
  - parent material (e.g. soils over the parent rock limestone)
  - topography (e.g. soils affected by slope (pp. 53–56))

Figure 3.5 **Importance of parent rock on soil type: podzol beneath heather on sandstone**

Ken Atkinson

- **Azonal soils** are those that are too young to have developed zonal characteristics (e.g. soils on sand dunes, recent river alluvium or fresh volcanic deposits).

Soil profiles and soil horizons

A **soil profile** is a vertical section through the soil down to its parent material. It is studied by digging a soil pit or by observing an exposed section in a road cutting or quarry. In such a soil profile there are roughly horizontal layers, or **soil horizons**, which can be distinguished by properties such as colour, organic matter, texture, structure and stoniness (Figure 3.6).

Figure 3.6 Diagram of the profile of an iron–humus podzol

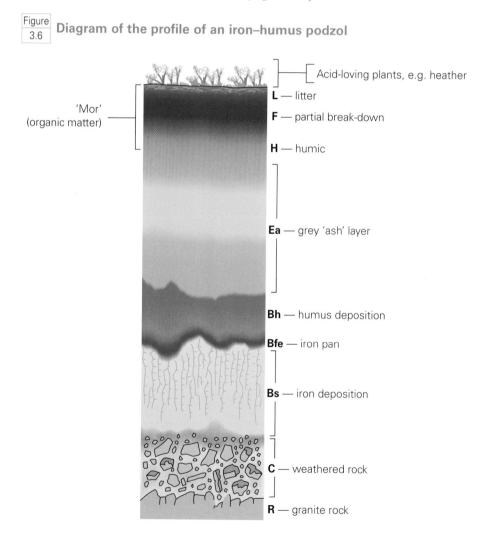

Acid-loving plants, e.g. heather

L — litter

F — partial break-down

H — humic

'Mor' (organic matter)

Ea — grey 'ash' layer

Bh — humus deposition

Bfe — iron pan

Bs — iron deposition

C — weathered rock

R — granite rock

The profile and its constituent horizons are the result of the soil-forming processes at work at the particular site. Thus, every soil has a sequence of horizons that can be identified according to processes operating within them. They are labelled using a letter scheme such as that shown in Table 3.2. The general horizons are indicated by upper case letters; detailed properties (suffixes) by lower case.

Table 3.2 Labels given to soil horizons

Horizons	Features
O	Peaty, wet organic surface
A	Intimate mixture of mineral and organic material
E	Severely depleted (**eluviated**) of chemicals and/or clay
B	Subsoil accumulation of organic matter and/or chemicals and/or clay
C	Unconsolidated parent material, such as glacial till
R	Bedrock parent material
Suffixes	**Features**
a	Ash coloured, e.g. Ea
b	Brown and mildly leached, e.g. Eb
ca	Calcium carbonate (lime), e.g. Bca
fe	Accumulation of iron oxides into a distinctive pan, e.g. Bfe
g	Mottled due to gleying, e.g. Bg
h	Humus, e.g. Ah
s	Iron and aluminium oxides, e.g. Bs
t	Accumulation of clay, e.g. Bt
w	Altered by weathering, e.g. Bw

Soil names and soil classification

Dokuchaev and his followers named soils according to Russian local names; some are still used today. Unfortunately, there is no universally agreed system of soil naming and soil classification. Different nations have devised systems that suit their own needs. In the UK there are three different schemes in England and Wales, Scotland and Northern Ireland! However, the picture is not wholly chaotic, because some names are common to different schemes. The main soil groups used by the Soil Survey of England and Wales are given in Table 3.3, together with their approximate percentage coverage.

| Table 3.3 | Soil groups in the *Soil Survey of England and Wales* |

Soil group	Main characteristics	Drainage	Cover (%)
Lithomorphic soils	Thin soils over solid rock	Rapid	7
Gleys	Signs of permanent waterlogging	Poor	40
Pelosols	High clay content leads to cracking when dry	Poor	10
Brown earths	Brown colour, leached and well-drained	Good	35
Podzols	Extremely acid, with raw humus; iron and aluminium moved from the Ea horizon to the B horizon	Good	5
Peat	Thick accumulation of organic matter	Poor	3
Human-made soils	Urban areas, composted soils, mines, archaeological sites	Mostly good	Minimal

International classifications

There are two widely used international schemes for naming and classifying soils:

- *Soil Taxonomy* was devised by the US Department of Agriculture in 1975 and is widely used. It employs a new scheme for naming soils, which explains partly why it has not been adopted universally.
- The Food and Agriculture Organization and the United Nations Educational, Scientific and Cultural Organization jointly published the *FAO–UNESCO Soil Map of the World* in 1974. It was revised in 1998 and retains a large number of traditional names.

Soil processes and profiles in well-drained situations

The importance of soil drainage in determining which soil-forming processes occur at a particular place is so important that it is convenient to divide soil types into those where drainage is free and unimpeded, and those where drainage is poor because of some factor in the soil or environment.

Podzols

Podzolisation involves thorough leaching of base cations (calcium, magnesium, potassium and sodium) from the soil under conditions of extreme soil acidity, brought about by a permeable, acidic parent material (e.g. glacial sands or weathered granite) and acidifying vegetation (e.g. heather or conifers). Below pH 4.0, rock minerals and clay minerals are weathered into constituent silica, iron compounds and aluminium compounds. The iron and aluminium compounds are removed from the surface Ea horizon, which is, therefore, ash-grey in colour, into an underlying red-brown Bs horizon (Figure. 3.7). Organic acids are produced by slowly decomposing surface 'humus' (**mor**), which is found in such situations. Mor has distinct litter, fermentation and humic layers (L–F–H). The organic acids react with the iron and aluminium to form chemical complexes called **chelates**, which allow the iron and aluminium to be transported.

| Figure 3.7 | **Distinct Ea and Bs soil horizons in a podzol on weathered granite in the Scottish Highlands** |

Ken Atkinson

Ken Atkinson

It is possible for podzols to occur under conditions of low rainfall. However, where the climate is wetter, there is often movement of organic matter from the surface into a subsoil Bh horizon above the Bs. If drainage is impeded in the subsoil for some reason, a thin pan-like accumulation of iron often occurs as a Bfe horizon (Figure 3.8). Podzols are typical of the boreal forest biome (Chapter 6). In Britain, they occur on upland heather moors and on unimproved land on acid rocks — for example, the granites and gneisses of Scotland, the Carboniferous sandstones of the Pennines and the Tertiary sands of the New Forest.

Figure 3.8 **An iron pan on weathered granite**

Ken Atkinson

Brown earths

Brown earths are the commonest soils in England and Wales. They occur on intermediate and basic rocks and on derived deposits, such as glacial tills. They are found in free-draining situations under conditions of continual leaching by rainwater of pH 5.5. The soils are only mildly acidic (pH 6.0) because of the recycling of nutrients leached into the subsoil by deep-rooting deciduous trees and by grasses; an action that works as a 'nutrient pump'. The surface humus (**mull**) is well humified, biologically extremely active, and is a well-structured, intimate mixture of organic and mineral materials.

Types of brown earth

There are two types of brown earth, recognised by the nature of the leaching:
- The first type is formed by the leaching of cations (calcium, magnesium, potassium and sodium) and nitrate anions in solution. Under undisturbed temperate deciduous forests, the typical soil profile is Ah → Eb → Bw → C or R. These soils are being universally cleared for farming, so the Ah, Eb and the top of the Bw horizon have been mixed and homogenised by ploughing.
- The second type is formed by **lessivage**, a leaching process that moves insoluble particles, for example clay minerals, down the soil profile in suspension. This type of brown earth is usually called a **luvisol** (derived from

eluviation — washing out). Lessivage is detected by thin coatings of clay, called **clay skins**, on the surfaces of soil aggregates in the subsoil. The sequence of soil horizons is Ah → Eb → Bt → C or R (Figure 3.9).

Figure 3.9 **Leached brown earth, with lessivage of clay from the pale horizon into the orange layer below**

Ken Atkinson

Grassland soils

The temperate grasslands of prairies, steppes and the pampas have the most fertile soils in the world. They are called **chernozems** in the FAO–UNESCO scheme, and **mollisols** in *Soil Taxonomy*. They have a thick, black Ah surface horizon, which is rich in organic matter and which stays crumbly when dry. This excellent structure allows easy penetration of roots and moisture. As they decay, grasses, with their fibrous roots, return a large amount of organic matter to the soil. Low soil-water levels during the warm summers slow down decomposition and allow humus to accumulate. Darkening normally occurs to a depth of 60 cm, with less intense colouring down to 1.5 m. The climate is subhumid to semi-arid, so leaching is weak and a horizon with calcium carbonate (Bca or Cca) is present at a depth of 1.0–1.5 metres (Figure 3.10).

| Figure 3.10 | Chernozem soil |

With such abundant additions of organic matter each year, it is not surprising that 75% of the biological activity of grasslands takes place in the soil. The large biomass of microflora (algae, fungi, bacteria and protozoa), microfauna (e.g. springtails, mites and ants) and soil mammals (e.g. gophers, prairie dogs and badgers) is estimated to turn over completely the top 60 cm of soil every 100 years. Nutrients are also released as the organic matter is decomposed and the weak leaching means that most of these stay in the soil. Grasses are efficient 'nutrient pumps', so the base cation content of the soil remains high and pH levels are correspondingly neutral or mildly alkaline.

Soil processes and profiles in poorly drained situations

Anaerobic effects

Approximately 50% of British soils are affected by either permanent or seasonal waterlogging. This produces **anaerobic** (oxygen-starved) conditions. Ferric iron (Fe^{3+}) in the mineral material is reduced to ferrous iron (Fe^{2+}), a process that is evident from the colour change from red to grey. This soil-forming process is called **gleying** and results in **gley soils**. In anaerobic environments, the activity of soil microorganisms is restricted. Undecomposed or partially decomposed organic matter accumulates as **peat**.

Gley soils

Gley soils have grey horizons that indicate chemical reduction. **Surface-water gleys** are caused by surface water being unable to drain through the soil because of its low permeability, perhaps due to a soil pan of compacted clay. A typical soil profile is O → Ah → Ebg → Btg → C. In **groundwater gleys**, it is the subsoil, rather than the surface, that is permanently waterlogged, due to a high water table. The horizons are O → Ah → Bg → Cg (Figure 3.11).

| Figure 3.11 | **Two types of gley soil** |

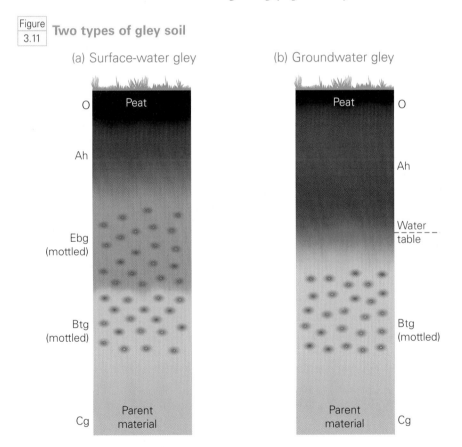

(a) Surface-water gley   (b) Groundwater gley

In addition to grey coloration, gleyed horizons have red mottles. These indicate where oxygen can penetrate — for example, along root channels and animal burrows, or around stones. Mottling can also denote seasonal gleying, with oxidation in the summer months and reduction in winter.

Peat soils

Peat soils occur where waterlogged conditions allow only a slow breakdown of plant remains. In uplands, where rainfall and humidity are high and the temperature is low, **blanket peats** (or **blanket bogs**) form, as in the Pennines, parts of Wales and the Lake District (Figure 3.12). These acidic environments, with sphagnum mosses, sedges and heather vegetation, favour acidic peats with a pH of less than 4.0. In contrast, **basin peats** occur in seepage sites that receive less acidic water and chemicals from surrounding catchments. The richest peat soils in the UK are in the East Anglian fens, where mineral-rich waters seep into lowland peats and silts to give soils with pH greater than 7.0.

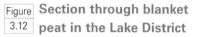

Figure 3.12 **Section through blanket peat in the Lake District**

Michael Raw

Lateritic soils

Lateritic soils (Latin, *later*, a brick) occur widely in tropical rainforest, seasonal forest and savanna biomes (Chapter 5). They are called **ferralsols** in the FAO–UNESCO scheme and **oxisols** in *Soil Taxonomy*. They were originally thought to result from weathering and leaching processes only, but it is now known that changes in the depth of the water table in seasonal forest and savanna biomes produce important gleying effects.

The soils result from a complex set of soil-forming processes. High temperature and high rainfall lead to intense

Figure 3.13 **Lateritic soil (oxisol)**

Ken Atkinson

chemical weathering of rock minerals and, therefore, deep soils. The base cations, calcium, magnesium, potassium and sodium, are thoroughly leached from the soil and there is loss of silica. This leads to a relative accumulation of the oxides of iron (haematite, Fe_2O_3) and aluminium (gibbsite, Al_2O_3). The latter is white but haematite is red and gives tropical soils their strong red colours. Despite the large biomass of tropical vegetation and the heavy litter-fall, the rapid rate of decomposition means that the surface humus horizon is thin.

The middle zone of laterites is distinctly mottled because of alternating oxidation and reduction with the seasonal change in the height of the water table. Below the middle layer is a pallid, almost white, zone where the conditions are permanently reducing. Under these conditions, ferrous ion (Fe^{2+}) is highly mobile and is removed. Therefore, a typical lateritic soil has five zones:

- thin surface humus
- a zone of iron and aluminium oxides
- a mottled zone
- a pallid zone
- deeply weathered rock

The soil catena

Mapping catenas

Topography is an important local soil-forming factor (p. 43), which influences intrazonal soils. When soil surveyors first began to map soils, characteristic sequences were recognised in the landscape on moving from the topographic divide or watershed, down the valley slope, to the valley stream, or to the stream bed in arid climates. This gradient of soil properties and soil types was found in widely separated places — for example, Africa, Canada and New Zealand. It is now recognised to be as fundamental as the soil profile in the study of soils.

In eastern Africa in the 1930s, Geoffrey Milne called the sequence a **soil catena** (Latin, *catena*, a chain) because individual soil profiles in the sequence are connected like the separate links in a chain. All regions on Earth have a characteristic soil catena — for example, the arctic tundra of Canada, coniferous forests in Scotland or tropical forests in Kenya.

Catena processes

The possible slope and soil processes that occur in a catena and that link the soil profiles in the chain are shown in Table 3.4. It is important to understand three principles:

- Not all catenas show all processes. For example, because of low temperatures, catenas in the Arctic show little chemical weathering or chemical leaching; gentle slopes do not show rapid mass movements.
- The catena model emphasises that soil and slope processes are inter-dependent. They cannot be separated when the **soil landscape** is studied in fieldwork or surveyed for maps.
- Milne thought of a catena as being on one uniform parent material, such as granite or a sandy glacial till. However, parent material usually changes down a catena because mass movement and erosion processes move material from one part of the slope to another.

Table 3.4 Processes occurring in soil catenas

Process	Process details	Process result
Physical and chemical weathering	Stable slopes; adequate moisture and temperature	Soil deepening due to soil formation
Mechanical movement of soil particles	Soil creep	Terracettes; movement of soil downslope
Soil erosion on slopes by water flow	Sheet, rill, gully erosion on slopes moves material downslope	Wind erosion on exposed sections of the slope
Movement of silts, loess	Rapid mass movements, (e.g. falls, slides, slumps, flows)	Transfer of material to lower slopes and river valley
Chemical leaching	Subsurface water action	Chemicals dissolved, trans-ported as ions and re-deposited in lower slope locations
Gleying	Chemical reduction	Gleyed soils
Organic matter accumulation	Minimal oxidation of organic matter due to waterlogging	Darker surface organic horizons, perhaps peats

A catena in a humid region

Figure 3.14 shows a catena in Scotland in which the role of water movement is paramount.

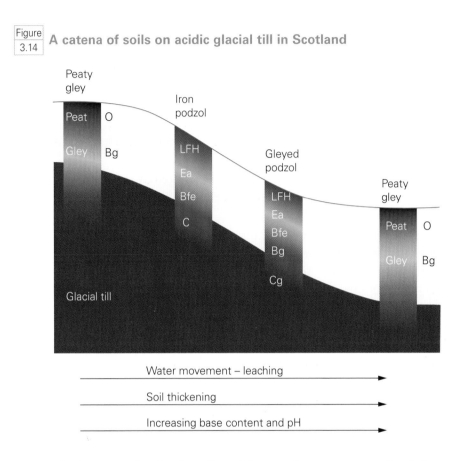

Figure 3.14 **A catena of soils on acidic glacial till in Scotland**

Poor drainage on the flat interfluve (the area between two valleys) favours gleying and peat formation (**peaty gley**); free drainage of water on the upper slope promotes leaching and podzolisation (**iron podzol**). Soils on lower-slopes show subsoil gleying as water is received from upslope (**gleyed podzol**). In the basin site, a dark peaty surface overlies a gleyed subsoil (peaty gley). Not only does water concentrate downslope, but it also leaches base cations to low-lying sites. The base cations, therefore, occur in higher concentrations in the basin peaty gley (> pH 5.5) than they do in the interfluve peaty gley (pH 4.5).

In general, catenas in humid environments lose minerals and chemicals from upper parts of the landscape (**shedding sites**) to the lower parts (**receiving sites**).

A catena in an arid region

Arid regions have little organic matter and only rare, but intense, rainfall. Here, catenas reflect the mechanical processes of mass movement and erosion. They have thin eroded soils on slope crests and deeper soils on colluvium, fans and alluvium in the lower part of the slope sequence.

Catenas in arid Australia reflect a complicated climate history. Figure 3.15 shows a catena in which the upper slopes are, in effect, subsoils of lateritic soils formed under more humid conditions. The onset of desert conditions in the Quaternary period hardened the mottled zone of the Tertiary laterite. The sequence of events over time is as follows:

- **Stage 1** The formation of lateritic soils in a humid tropical climate.
- **Stage 2** Climate change to drought conditions, with erosion of the surface horizon exposing the mottled zone.
- **Stage 3** Dehydration of iron and aluminium oxides in the mottled zone produces a hardpan at the surface (**cap-rock**).
- **Stage 4** Mineral material is mechanically transported into the adjacent valley.
- **Stage 5** Continuous build-up of mineral material in the valley (**aggradation**) and continuous burial of valley soils (**buried soils**).

| Figure 3.15 | **A catena of soils in the Australian desert** |

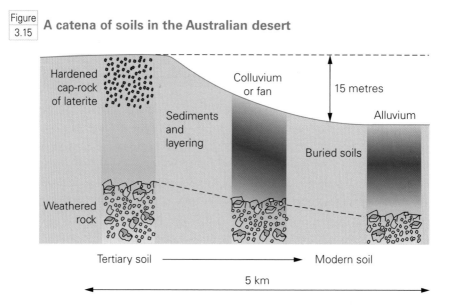

Just as the movement of surface and subsurface waters provides the links between the different soils in the Scottish catena, so past chemical, and past and present mechanical processes, link the soils in the Australian catena.

4 Soils: modification, conservation and sustainable use

Effects of farming on soils

Agriculture has changed over the last 50 years, in what has been called the 'second agricultural revolution'. The result is larger farms, use of fertilisers and pesticides, more intensive and specialised cropping and larger and heavier machinery. In the 1980s, large increases in yields in the UK produced an excess of production over consumption, a situation not seen for several centuries. Surpluses of food and increasing concern about environmental pollution from farming led to:

- diversification into alternative crops and livestock
- increased environmental controls, as in Environmentally Sensitive Areas, Nitrate Vulnerable Zones and Countryside Stewardship
- increased interest in organic and biodynamic farming systems

Agriculture is a complex industry that is involved in different enterprises and uses a range of operations. Table 4.1 summarises some of these operations in the UK and records the likely effects on soil properties. A soil profile in an arable field shows a ploughed surface horizon (Ap) to the depth of ploughing at 20 cm (Figure 4.1). In pasture, the surface horizon is an organic-matter-rich Ah, reflecting the effects of the grasses on the soil profile (Figure 4.2)

The influence of cultivation on soils depends upon specific conditions at the time. For example, the effects of ploughing a clay soil depend upon its moisture level. If ploughed when too dry, it breaks into large clods that roots cannot penetrate; if ploughed when too wet, it smears and compacts to form a 'plough pan'. If fertilisers are applied at the wrong time, in the wrong chemical form or the wrong amount, there are risks of losses in surface runoff and leaching, and of polluting streams and lakes.

Table 4.1 Effects of farming operations on soil properties

Operation	Soil property	Effect
Ploughing	Horizons	The A and the top of the B horizon are homogenised into a 20 cm thick Ap horizon
	Structure	Larger units or 'clods'
	Air content	Increased
	Infiltration	Increased
	Organic matter	Decreased due to oxidation and crop removal
Manuring	Colour	Darkened
	Organic matter	Increased
	Structure	More stable
	Nutrients	Increase in all nutrients, particularly nitrogen and phosphorus
	Microorganisms	Increased activity
Use of fertilisers	Nutrient content	Increased
	Leaching loss	Increased
Use of pesticides	Pollution	Build-up of non-biodegradable chemicals
Drainage	Gleying	Water table lowered, giving greater rooting depth
Irrigation	Water content	Increased
	Salinity	Increased risk, especially if fertilisers used
Crop rotation	Fertility	Increased if legumes, livestock and root crops are rotated with cereals or oilseed rape
Liming	Acidity	Reduced
Temporary grass growing	Colour	Darkened
	Organic matter	Increased by root biomass
	Nutrients	Increased, especially nitrogen and phosphorus
	Structure	Improved, structures (aggregates) stable when waterlogged
Use of machinery	Compaction	Increased
	Infiltration capacity	Decreased

Figure 4.1 **Soil profile in a cultivated field**

The light-brown surface horizon shows the depth of ploughing

Figure 4.2 **Soil profile in a pasture**

The dark-brown surface horizon shows organic matter derived from grass roots

Sustainability

The term **sustainability** has recently come into common usage. However, the concepts behind it are old, going back to prehistoric and biblical times. One meaning of sustainability is that production does not harm the 'resource base', i.e. crops are produced without lowering — and hopefully increasing — the capacity of the soil to produce them. A second meaning emphasises the economic viability of the production unit, i.e. the farm can continue indefinitely to maintain profitability. In reality, there is a case for including both these aspects in the meaning of the term.

Sustainable agriculture is agriculture that incorporates certain guidelines:

- good soil conservation that also achieves good water conservation
- feeding the soil by recycling nutrients and maintaining soil conditions conducive to growth
- rotating different crops to limit weeds, pests and diseases
- reduced and timely cultivations (**minimum tillage** or **zero tillage**)
- integrating livestock to use crop by-products and to provide manures
- use of human waste with appropriate care and safety
- sharing of experience and knowledge by local farmers for the benefit of all

Effects of forestry on soils

Some of the Earth's remaining natural forests are found on acid soils — for example, the boreal forests (Chapter 6) and the acid oak woods of western Europe. Since the 1920s in Britain, particularly in Scotland and Wales, there has been widespread planting of monocultures of exotic conifers as part of national forestry policy (Figure 4.3). This **afforestation** leads to changes in soil acidity and to the chemistry and sediment load of streams.

Figure 4.3 **Coniferous plantation of Sitka spruce, Ennerdale, Lake District**

Ken Atkinson

Soil acidity

Coniferous plantations cause soil acidification, especially in the top 20 cm of the soil profile. The reasons for increased acidity are:

- the acidic mor humus on the surface derived from conifer needles and twigs
- organic acids, which leach nutrients more effectively
- podzolisation
- the removal of base ions such as calcium and magnesium from the soil by roots and their 'locking up' in the trees
- the removal of trees with these nutrients in the timber

Wide variations in the degree of acidification by plantations have been reported, with reductions ranging from 0.2 to 2.0 pH units — this is equivalent to about a 60-fold variation in hydrogen ion concentration (pH is a logarithmic scale). The long-term impact depends on the species of tree, the initial soil conditions, the forestry practices used and the proportion of the tree removed in harvesting.

The removal of trees can also cause soil acidification. Dimbleby studied the effects on soils of the clearing of deciduous forests on the North York Moors (northern England) by Bronze Age and Neolithic farmers. By examining soils under prehistoric burial mounds, he showed that clearance led to an invasion of heather, which resulted in extremely acidic podzolic soils. He also showed that this human-induced acidification is reversible — planting the heathland with birch and oak can convert podzols back to brown earths that have both higher pH and higher base ion contents.

Stream chemistry

Coniferous plantations lower the pH of streams. This impacts adversely on the fish population, particularly that of trout. It is a serious problem in areas with acidic geology in Scotland, Wales and northern England. Part of the problem is that plantations appear to promote surface runoff at the expense of subsoil drainage. Subsoil drainage increases the chance of acidic water being partially neutralised by passage through the soil. Subsoils have higher pH values than surface litters. For this reason, forest drains on the surface should end well short of streams in order to promote drainage through the deeper mineral soil.

Felling forests increases the nitrate content of the soil. There are large leaching losses of nitrate to streams because nitrate cannot be held by soil colloids and it is no longer being taken up by trees (Chapter 1). This results in pollution of streams and loss of nutrients from the ecosystem. This was shown in the Hubbard Brook experiment in the USA, where an entire watershed was clear-felled and changes in soil and water chemistry were recorded. In North Wales, losses of $70\,kg$ nitrogen $ha^{-1}\,yr^{-1}$ have been found after the felling of a Sitka spruce plantation. This is eight times higher than nitrogen loss before felling.

Sediment load

In upland Britain, forestry usually involves the drainage of poorly drained peats and peaty gley soils by a network of deep ditches called **grips**. Gripping results in more rapid and greater surface runoff and brings an increased risk of soil erosion and flash flooding. More 'flashy' rainstorms (rainfall that is more intense and convectional in nature) are one of the predictions of climate change research, so flooding could become more serious. Streams draining from catchments in the early stages of afforestation projects have a large suspended sediment load, in contrast to the clear streams with little load coming from catchments with mature forests (Figure 4.4).

Figure 4.4 | **Sediment in the lower stream is from a catchment being prepared for forestry**

Ken Atkinson

Maintaining soil fertility in conventional, conservation and organic farming systems

Within the past 60 years, farming in the UK has undergone radical change. From 1945 to 1955, there was widespread mechanisation of farming; between 1960 and 1980, conventional agriculture became dependent on agrochemical inputs of fertilisers and pesticides. Concerns about soil compaction, environmental pollution and overproduction led to the search for alternative systems to make agriculture more environmentally friendly. Since 1995, purely 'chemical farming' has declined relative to 'conservation farming' or 'low-input systems'. Systems based on organic matter have become more popular (Table 4.2).

Table 4.2 A classification of farming systems

Farming system	Type
Conventional systems	High-input agriculture or 'chemical farming'
Alternative systems	Conservation or low-input agriculture
	Organic agriculture
	Biodynamic agriculture

- **Conservation agriculture** emphasises minimal cultivation, organic manure, crop rotation and mixed cropping to reduce pests, but also uses inorganic fertilisers and minimal pesticides.
- **Organic agriculture** uses no synthetic fertilisers or pesticides. It concentrates on:
  – building up soil fertility by adding manure and compost
  – controlling pests by crop rotation and by crop and livestock diversification
  Most organic farms are certified by the Soil Association.
- **Biodynamic farmers** follow organic principles, but farm in a holistic way and in empathy with the cosmos. They add natural preparations made from materials such as cow manure, dandelions and nettles to improve the quality of crops, soils and manures. They are certified by the Biodynamic Association.

Comparisons of soil quality

When comparing soil quality in conventional farming with that in organic farming, it is important to compare specific land uses — for example, wheat with wheat and vegetables with vegetables; comparisons of whole systems would not necessarily compare like with like. Organic farms have better soil quality than neighbouring conventional farms. They have larger amounts of organic matter, more active soil microorganisms (especially soil mycorrhizae — Chapter 5), more earthworms, better soil structure, lower bulk density, easier soil penetrability and thicker topsoil.

Differences in chemical properties are less clear-cut. Nitrogen content is often higher on organic farms; other nutrient levels may be higher on conventional farms, as might be expected from adding chemical fertilisers. Soil pH is often lower on organic farms because decomposition of organic matter in the soil releases hydrogen ions.

It is mostly in physical and biological properties that soils on organic farms are superior. Some people believe that soils on organic farms will withstand climate change better than those on conventional farms. This is because plant roots penetrate more deeply into the soil in organic systems and can, therefore,

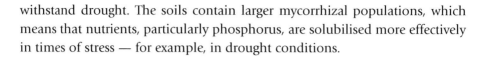
withstand drought. The soils contain larger mycorrhizal populations, which means that nutrients, particularly phosphorus, are solubilised more effectively in times of stress — for example, in drought conditions.

Soil acidification

The burning of fossil fuels, wood and waste produces emissions rich in sulphur (sulphates, SO_4^{2-}; sulphur dioxide, SO_2) and nitrogen (nitrogen dioxide, NO_2; nitrous oxide, N_2O; nitric oxide NO). These form sulphuric acid (H_2SO_4) and nitric acid (HNO_3) in the atmosphere. Hydrochloric acid (HCl) is produced by the burning of municipal waste. The pH of natural precipitation is 5.5, but precipitation with pH less than 3.5 has been found in heavily industrialised regions. **Acid rain** is strictly **acidic precipitation** because there is much acidity in snowfall, especially with increased power generation and use of central heating in the winter months.

Using fertilisers in agriculture produces soils that are more acidic because nitrates and sulphates are commonly used. The world's oldest agricultural experiment is the Park Grass Experiment, started in 1856 at Rothamsted Experimental Station, Hertfordshire. The soil was originally pH 5.9. In the past 150 years, acidity has increased to pH 3.9 on plots receiving fertilisers but no lime. It is estimated that in the 1850s, the amount of nitrogen received by the soil in precipitation was $10\,\text{kg}\,\text{ha}^{-1}\,\text{yr}^{-1}$, compared with $45\,\text{kg}\,\text{ha}^{-1}\,\text{yr}^{-1}$ today.

This is a significant increase in the addition of nitrogen — a major nutrient — to ecosystems and crops in Britain. However, soil acidification has a number of adverse effects. It increases the rate of leaching of bases and increases the solubility of heavy metals. At Rothamsted, hay from unlimed 150-year-old plots has 40 times the quantity of aluminium ions compared with hay from limed plots. It is now seven times higher than the level tolerated by cattle. The activities of soil microorganisms in decomposition and cycling of nutrients have been reduced greatly. Direct damage to the leaves of trees, giving 'dieback', has become a serious problem throughout Europe.

Soil erosion

Soil erosion is the accelerated removal of soil at rates greater than those of natural erosion. It is usually due to poor land management. The adverse effects of soil erosion are:

- removal of soil, humus and nutrients, which are concentrated at the surface
- thinning of soils, hence giving less depth for rooting and water storage

- increased sediment load of streams, leading to increased risk of flooding, and the silting up of drains, irrigation channels and reservoirs
- **eutrophication** and pollution of lakes and streams by sediment load containing nutrients and farm pesticides

The US Soil Conservation Service (USSCS) gives some 'ballpark' figures for soil erosion on the global scale. Assuming soil formation of 1 tonne ha^{-1} yr^{-1}, it sets the maximum acceptable loss in the USA at 5 tonnes ha^{-1} yr^{-1}. Extreme losses can reach 30 tonnes ha^{-1} yr^{-1} from cropland in Africa, Asia and Latin America. Losses of 17 tonnes ha^{-1} yr^{-1} have been recorded in the Mediterranean region — one of the worst affected areas in Europe (Figure 4.5). In experiments on the UK South Downs, soil losses of 26 m^{3} ha^{-1} yr^{-1} were recorded on heavily eroded fields (pp. 67–68).

Figure 4.5 **Severe soil erosion in Mediterranean Spain**

Ken Atkinson

Types of soil erosion

Rainsplash breaks down aggregates on the soil surface by the impact of the falling raindrops. On a slope, there is net transport of the material downwards. **Soil wash** is the term for this type of erosion. Transport by running water is the chief means by which soil particles are transferred from their original position. Surface runoff is initiated when rainfall intensity exceeds the infiltration capacity of the soil. Erosion risk is increased when infiltration capacity is lowered by surface compaction, which can be caused by raindrop impact producing a surface cap, by pressure from machinery and, on paths, by human feet.

Once overland flow has started, it can traverse the ground as:

- a uniform sheet, giving **sheet erosion**
- a small channel or rill, giving **rill erosion**
- a deeper channel or gully, giving **gully erosion** (Figure 4.6)

Figure 4.6 **Soil erosion by gullying**

Ken Atkinson

There is controversy over sheet erosion and sheet overland flow. Some experts think that it is rare, as overland flow quickly channels itself into depressions where flow is faster and turbulent. Rills are transient features that are usually destroyed by ploughing.

The intensity of soil erosion, usually called the **erosion risk**, depends upon two key environmental conditions:

- the **erosivity** of the agent of erosion
- the **erodibility of the soil**

Erosivity of the eroding agent

Erosivity refers to the energy of the rainfall or the wind and its ability to move soil particles. As with all agents of erosion, the work done depends on an **intensity factor** and a **duration factor**. The most important protective factors that lessen the impact are vegetation and surface stoniness.

Soil erodibility

Soil erodibility is a soil property that describes the ease with which the surface structures of a soil are broken down, and the particles transported out of a field or ecosystem, or deposited at the base of a slope. It depends upon:

- soil texture
- soil structure
- slope length
- slope steepness
- field relief
- slope profile and convexo–concave shape

Texture governs whether the soil is prone to slaking, disaggregation and capping (crusting) under the effects of waterlogging or powerful raindrop impact. Silty textures (silty clay loams and silt loams) are susceptible to capping (Figure 4.7).

| Figure 4.7 | **Soil surface sealed by 'capping'** |

(a) Energy impact of heavy rainfall breaks down unstable aggregates

(b) Soil particles, released into suspension, settle and block pores

(c) Soil crust reduces infiltration rate to zero. Water in puddles, with high
losses by evaporation and runoff

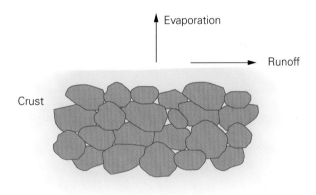

Structural stability relies on organic matter to provide bonds in the clay
domains (Chapter 3). Highest erosion risk occurs with a slope length greater
than 200 m, maximum slope angle greater than 10°, relief within a field greater
than 30 m and with convex slopes and wide slope crests.

Land use and erosion risk in the South Downs

Geographer John Boardman has kept a log of soil erosion events in arable fields
on the chalk South Downs of England since the 1970s. The soils are shallow
rendzinas (<0.3 m depth to rock) and are silt loams with stones and flints. Most
soil is moved in rills and gullies, with little sheet erosion. In the 1970s, wet
autumns gave little erosion. The risk of erosion increased during the 1980s
because of a switch from the growing of spring cereals to winter cereals, which

meant that the presence of large areas of bare land coincided with the maximum rainfall in autumn. Spring cultivation allows a protective stubble during winter. About 70% of erosion events were in winter-cereal fields, 10% in grassland, and 17% in other cultivated land.

This intensification of arable farming was nationwide. It included the creation of larger fields, the cultivation of steeper slopes, compaction by farm machinery and cultivation down slopes rather than along the contours. Intensification has also decreased the organic matter content of soils, which makes it difficult to maintain stable structures. Perhaps most serious, however, has been the production of a fine, easily eroded seedbed for the winter cereals, by drilling and rolling in early autumn.

Rainfall and erosion risk in the South Downs

The erosivity of the rainfall is also a factor in soil erosion on the South Downs. The highest erosion risk occurs when there are several erosive rainfalls at regular intervals during the autumn, which produce slaking and capping. These are then followed by heavy rain during the winter months, leading to the formation of major rills. If rainfall events become more concentrated and 'flashy' because of global warming, rainfall erosivity is set to increase (p. 65).

Activity 1

The table below shows erosion rates and physical features for four ploughed/cultivated fields monitored by Boardman over a 12-month period.

Field	Maximum angle (°)	Slope length (m)	Relief (m)	Erosion rate (m³ ha⁻¹)
1	15	560	53	13
2	19	260	50	26
3	10	200	20	23
4	12	320	45	12

(a) Choose suitable graphical methods and construct graphs to display these data.

(b) Describe and explain the effects of the physical features on erosion rates.

(c) Discuss the relative erosion risks in British farming of:

    (i) winter oilseed rape

    (ii) spring barley

    (iii) maize

Activity 2

Study Figure 4.8, which shows calculated rates of erosion and aggradation for two fields, superimposed on isometric projections. The site in Sussex is on chalk; the site in Shropshire is on sandy soil.

(a) For each field, describe the pattern of erosion and aggradation.

(b) Describe and explain the main differences in erosion and deposition between the two fields.

Figure 4.8 **Distribution of erosion and aggradation across two fields**

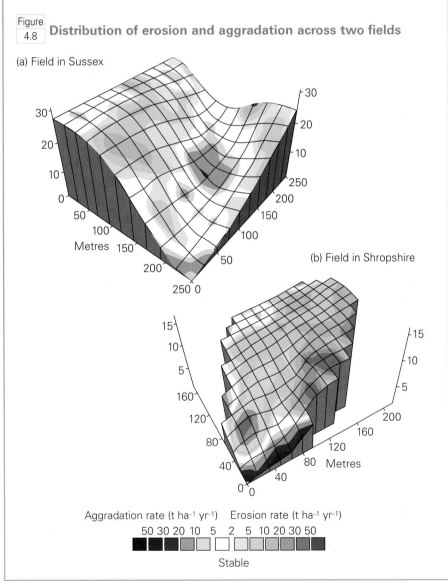

(a) Field in Sussex

(b) Field in Shropshire

Aggradation rate (t ha⁻¹ yr⁻¹)   Erosion rate (t ha⁻¹ yr⁻¹)
50 30 20 10 5   2   5 10 20 30 50

Stable

5 Biomes: properties, distribution and tropical forests

Properties of biomes

Physiognomy and structure

Biogeographers have always been interested in differences in vegetation and wildlife across the Earth. During early voyages of exploration, for example those of Captain Cook in the eighteenth century and of Charles Darwin in HMS *Beagle* in 1831–36, the natural history of the globe was studied in a scientific way. It soon became clear that vegetation formations with similar plant physiognomy and vegetation structure occurred on different continents, often thousands of miles apart and not sharing any of the same species.

Plant physiognomy refers to the growth form, shape and function of plants. There have been a number of classifications of plants based on physiognomy. For example, that of the Canadian biogeographer Dansereau has six classes:

- trees
- shrubs
- herbs, including grasses
- bryophytes (non-flowering plants)
- epiphytes (grow on other plants)
- lianas (woody vines)

Vegetation structure refers to the total assemblage of plants in a vegetation community — for example, forest, woodland, shrubland, scrub, grassland and desert. Further subdivisions can be made on the basis of size, leaf shape and plant function.

Vegetation formations

A **vegetation formation type** is a large-scale region that has a uniform vegetation structure. Thus, grasslands in temperate climates, for example the **prairies** of North America and the **steppes** of Russia, are similar and homogeneous. They belong to one formation type. This scale of study was used when the Earth was being explored and mapped for the first time. It has become popular again in recent years and is being used by climate and vegetation modellers working at the global scale to predict changes in vegetation that will result from climate change.

Biomes

In 1936, two American biogeographers, Clements and Shelford, coined the term **biome** to describe large areas of the Earth that have similar climate, soils and vegetation. Like the vegetation formation type, the biome is a large-scale region. However, it refers to the total ecology of a region, i.e. climate, soils and wildlife, as well as plants. Hence, it is a more ecological term than formation-type. Figure 5.1 shows the terms used at decreasing scales in the study of plants and ecosystems.

| Figure 5.1 | **Units of study of plants and ecosystems at decreasing scales** |

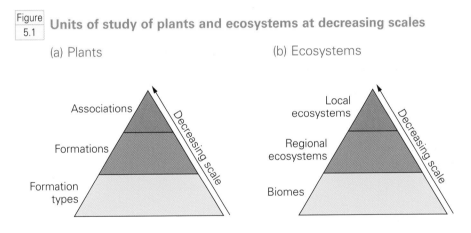

(a) Plants

(b) Ecosystems

The importance of climate in influencing vegetation was underlined, also in 1936, when the German climatologist Köppen produced a map of world climate regions that used vegetation boundaries (e.g. the position of tree lines and the limits of deserts) to locate climate boundaries. It was recognised that similar regional climates across the Earth produce similar plant physiognomy and vegetation structure; this is true even when the plant species are unrelated and when they are located on different continents, separated by large distances. Thus, the tropical rainforest of the Amazon is similar in plant growth form and overall structure to the tropical rainforest of Queensland, Australia, even though the plant genera and species are different. The two vegetation regions are called **ecological equivalents**; the process that produced them is termed **convergent evolution**.

The global distribution of biomes

Whittaker's classification

Several classifications of biomes have been devised. They are all similar in using some index of heat (e.g. temperature or potential evapotranspiration) and some index of moisture (e.g. precipitation or soil-moisture deficit). The classification of the American biogeographer Whittaker is simple and is used widely. It correlates global biomes with mean annual precipitation and mean annual temperature. However, it does not use any index of seasonal variation, which is a limitation because seasonality of climate is important. Whittaker divides the Earth into ten biomes ranging from tropical forests to Arctic–alpine tundra (Figure 5.2).

Figure 5.2 **Whittaker's classification of biomes**

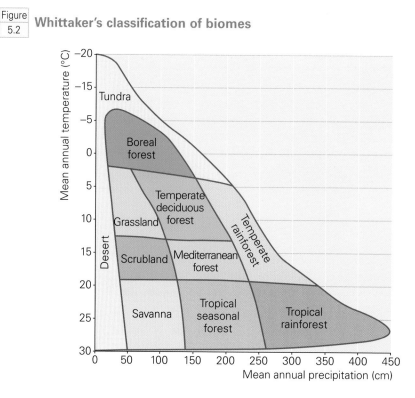

In Whittaker's model the main controls are climate, including the effects of mountain ranges in making climates cooler with altitude and in causing rain-shadows. Geographical location and environmental history, including use by humans, are now recognised as important factors in explaining the differences *between* and *within* biomes. Any map of world biomes is a map of 'potential vegetation' rather than a map of 'actual vegetation'. Little 'natural vegetation' still exists and few areas on the Earth are unaffected by human impacts. It has been estimated that 40% of annual plant growth is used in some way by humans.

Productivity and biomass

Table 5.1 gives estimates of area, net primary productivity (NPP) and total biomass for the 10 biomes in 1950, before changes were brought about by widespread clearance. There are differences in NPP ranging from 20 tonnes ha^{-1} yr^{-1} for tropical rainforests to 0.7 for hot deserts and 1.5 for polar tundra. Both NPP and biomass show an inverse relationship with the severity of limiting factors of cold and moisture-deficit in the different biomes; the greater the limitations, the lower the productivity.

Table 5.1   Net primary productivity (NPP) and biomass values for world biomes

Biome	Area (million km^2)	Net primary productivity (tonnes ha^{-1} yr^{-1})	Biomass (kg m^{-2})
Tropical rainforest	18	20	44
Tropical seasonal forest	8	5	36
Tropical savanna	15	7	4
Desert	26	0.7	0.67
Mediterranean forest	8	6	6.8
Temperate grassland	9	5	1.6
Temperate deciduous forest	7	10	30
Temperate rainforest	5	13	36
Boreal forest	12	8	20
Tundra	8	1.5	0.67

Tropical forests: nature, exploitation and sustainable management

Types of tropical forest

Tropical forests are of two types:
- **tropical rainforest**
- **tropical seasonal forest** (sometimes called **monsoon forest** in Asia)

Climate is the main factor producing the different types. Tropical rainforest has an equatorial **perhumid** climate, with high temperatures and moisture all year round, whereas tropical seasonal forest has similar high temperatures but a marked dry season in the winter months (Table 5.2).

Table 5.2 Key features of tropical rainforests and tropical seasonal forests

Feature	Tropical rainforest	Tropical seasonal forest
Global area	18 million km^2	8 million km^2
Position	Equatorial, 25°N–25°S	North and south of tropical rainforest
Temperature	Annual average >20°C	Annual average >20°C
Rainfall	Annual average 200–400 cm	Annual average 150–250 cm
Rainfall distribution	Little variation in the daily and seasonal pattern	Strongly seasonal monsoonal pattern, with dry winters lasting 4–7 months
Structure	Five layers or strata: • supercanopy of emergent trees 40 m • canopy trees 25 m • subdominant trees 10 m • small trees and tree ferns • sparse understorey of low shrubs and herbs	Three layers or strata: • canopy trees 25 m • shrubs and saplings • dense ground cover
Plant characteristics	Evergreen trees with leathery leaves; many vines, lianas and epiphytes; pollination and seed dispersal by insects, animals, birds and bats	Deciduous trees, which lose leaves when dry; shrubs, vines, herbs dormant in dry season; fewer lianas and epiphytes; fruit and seed dispersal in the dry season
Light penetration	Only 1% of radiation for photosynthesis reaches the forest floor	More radiation reaches floor, resulting in a dense ground covering of plants
Soil	Weathered, leached red laterites (oxisols); very low in plant nutrients, except where alluvial or volcanic	Weathered leached red laterites (oxisols); less leached than in tropical rainforest, so contain more plant nutrients
Biodiversity	Highest diversities of plants, insects, birds and some mammals (e.g. primates, bats) in the world	Similar diversities of insects and other animals, but plant diversity <50% of that of tropical rainforest

Physiognomy and structure

Tropical forests are found in Central and South America, Africa, south and southeast Asia, and eastern Australia. Similar general features can be seen in all these regions. However, there is no typical tropical rainforest or tropical seasonal forest because there are regional variations in the species of plants and animals and in the environmental and human histories (Figure 5.3).

<table>
<tr><td>Figure
5.3</td><td>**Tropical rainforest at 1500 m altitude in Sri Lanka**</td></tr>
</table>

Richard Smith

Tropical rainforest has plentiful heat and soil moisture, so the vegetation reflects intense competition for light. Tall, slender, evergreen trees make up 70% of plant species. Some have **buttresses** or **stilt roots** for extra support. The high relative humidity within the forest allows **epiphytes** and **vines** to thrive. Most are benign, although the strangler fig eventually kills its host. Leaves often have **drip tips** to allow moisture to drain more easily from their surfaces (Figure 5.4).

Figure **Adaptations in tropical rainforest plants**
5.4

(a) Branched crown and buttresses

(c) Pinnate leaves

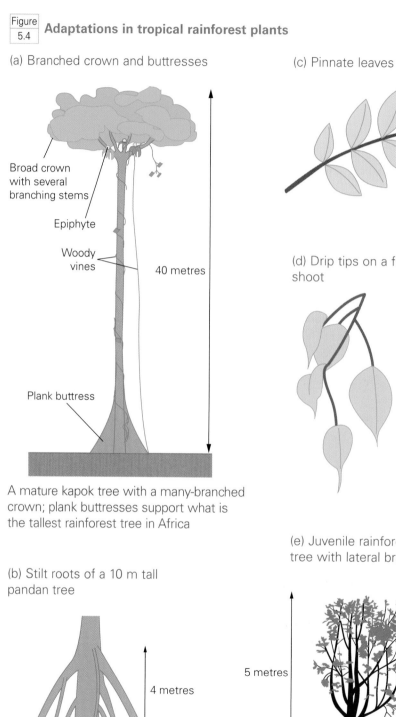

Broad crown
with several
branching stems

Epiphyte

Woody
vines

40 metres

Plank buttress

A mature kapok tree with a many-branched
crown; plank buttresses support what is
the tallest rainforest tree in Africa

(b) Stilt roots of a 10 m tall
pandan tree

4 metres

(d) Drip tips on a fig
shoot

(e) Juvenile rainforest
tree with lateral branches

5 metres

Tropical rainforest has a complex five-layer structure (Figure 5.5).

| Figure 5.5 | **Tropical rainforest structure and strata** |

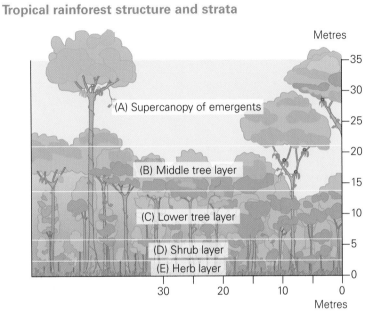

Tropical seasonal forest has a simpler three-layer structure, with plants showing xerophytic adaptations to the seasonal drought (Table 5.2). Its biomass and biodiversity are second only to tropical rainforest, although it does not receive as much media publicity and conservation interest.

Activity 1

Study this sketch of an area of tropical rainforest.

(a) Identify and name the strata in the vegetation.

(b) Briefly describe each stratum.

(c) Explain why there is well-marked layering, paying attention to the different environmental conditions in each layer.

Plant–insect interactions

Tropical rainforest contains 40% of all the plant species on Earth and 400 000 species of insect. Insects interact with the plants in complex ways. The American biogeographer Janzen has shown how tropical figs are dependent for pollination on a mutualistic relationship with wasps. Every species of fig has its own type of wasp. If a fig tree grows outside the range of its pollinator, fertile seeds do not develop. This relationship is termed mutualism, i.e. one in which both species benefit (Chapter 2).

Many species of insect interact with plants by feeding on them in parasitic relationships (one organism benefits at the expense of the other). Plants defend themselves by producing chemicals to counteract grazing. Some of these, for example quinine, are used as medicines.

Plant–insect interactions also help to explain the diversity of trees in tropical forests. According to Janzen's 'enemies hypothesis', tree seedlings have little chance of surviving near the parent tree because of attack by insects feeding on the parent tree; only those seeds dispersed far away by birds and animals will survive.

There is still much to learn about plant–insect interactions and about natural medicines.

Nutrition of tropical trees

Tropical soils are laterites (oxisols). They are infertile, leached and acidic. Therefore, they are poor in nutrients (Figure 5.6).

However, although the surface humus is thin, rates of decay and mineralisation of leaf-fall and litter are rapid at the soil surface. The dense network of fine roots allows nutrients to be captured quickly by the living vegetation (Figure 5.7).

Figure 5.6 Thin humus layer at the top of a tropical lateritic soil (oxisol)

Ken Atkinson

Figure
5.7 **Rapid and efficient nutrient cycling in tropical forests**

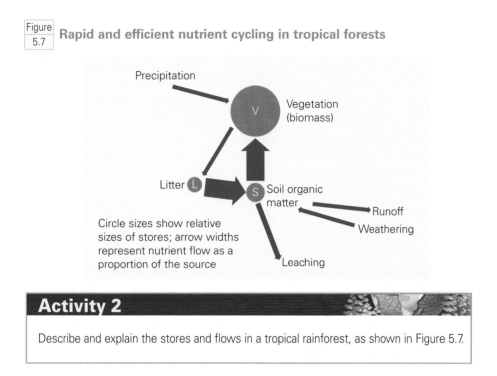

Activity 2

Describe and explain the stores and flows in a tropical rainforest, as shown in Figure 5.7.

Most trees have **mycorrhizae** (mutualistic 'root fungi') that penetrate and infect their root systems. The tree provides the fungus with carbon; the fungus absorbs nutrients from the soil and passes them to the host. Clearance of trees for farming, whether in shifting cultivation or for commercial agriculture, destroys both the root systems and the fungi. Shifting cultivation circumvents this because the plot is abandoned after about 5 years. This allows regeneration of trees and the nutrient-conserving mechanisms (Figure 5.8). Commercial agriculture has to use large quantities of fertilisers and manures to prevent the soil quality from declining.

Figure
5.8 **Shifting cultivation in the Sri Lankan rainforest**

Richard Smith

Tropical mountain vegetation

The difference in temperature between lowlands and uplands is greater in the tropics than at mid and high latitudes. Cooler mountains are common in all continents and islands in the tropics, and lowland biomes change quickly with altitude to **tropical montane forests**. These forests have smaller, 10 m high trees, with no supercanopy of emergent trees. Trees have smaller leaves, abundant epiphytes and no lianas or buttresses. Ferns and mosses cover the trees.

In the past, in some parts of the tropics, mountains were important for maintaining biodiversity. In Pleistocene glacial times, the tropics were drier than they are today. Plants and animals retreated upwards into wetter mountain refuges (**tropical refugia**) where they not only survived but evolved by natural selection, producing new **endemic species**. Endemic means 'confined to a small region and found naturally nowhere else'. Tropical mountains contain most of the Earth's endemic species.

The theory of tropical refugia is controversial. It is not thought to be relevant to South America, where the Andes in the west form a single highland range. It may be more relevant to tropical Africa (Figure 5.9) and Asia, where distinct and isolated mountains are widely distributed. However, biodiversity is greater in tropical South America than in tropical Africa. This illustrates the danger of making generalisations about a biome when there is still much to discover.

| Figure 5.9 | **Pleistocene refugia in the African rainforest** |

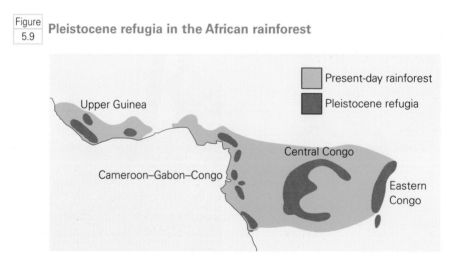

Human impacts and deforestation

Human activity poses a serious threat to tropical forests because growing populations and poverty in tropical countries lead to continual clearance and exploitation (Figure 5.10).

Figure 5.10 **Clearance of Sri Lankan rainforest for tea plantations**

Forests are regarded as resources to be used rather than to be preserved (Table 5.3). Few areas are without evidence of human intervention, and **secondary** tropical rainforest and tropical seasonal forests are now much more common than **primary** forests (Fig. 5.11).

Table 5.3 Human impacts on tropical forests

Land use	Impacts
Hunting and gathering	Little impact by the small bands of indigenous people
Traditional agriculture	**Shifting cultivation** (swidden, with local names such as *milpa* in Central America and *chena* in Sri Lanka) only supports a small population, yet the farming is in sympathy with forest ecology; **permanent gardens** for maize or root crops (e.g. yam, manioc, sweet potato) need fertilisers to maintain crop yields and to be sustainable
Commercial farming	Livestock ranching leads to poor quality grassland; paddy lands for rice sustain huge populations; soya farming requires massive pesticide use
Commercial plantations	Bananas, cocoa, coffee, oil palm, rubber, tea

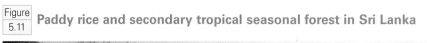

Land use	Impacts
Trophy hunting, plant and animal collecting	Severe impacts on prized species of mammals, birds, reptiles, fish, butterflies, orchids, bromeliads, ferns
Logging	Tree removal, road construction and general infrastructure; removing 10% of the trees destroys 60% of the forest
Reforestation	Replanting cleared forests with pine and eucalyptus has mostly failed because it leads to drying out of soils

Figure
5.11 **Paddy rice and secondary tropical seasonal forest in Sri Lanka**

Richard Smith

It is estimated that 170 000 km² of tropical rainforest, equivalent to 1% of the biome, are cleared each year, of which 45% is in the Amazon. The tropical seasonal forest is equally vulnerable as it is more easily cleared, particularly by fire in the dry season. Its natural regeneration is slower than that of the tropical rainforest.

Activity 3

Study Figure 5.12 (a) and (b), which show the extent of tropical rainforest in Sumatra in 1932 and 1985 respectively. With the help of data on the human environment given in Figure 5.12 (c), discuss the nature, causes and distribution of deforestation on the island.

Activity 3 (continued)

Figure 5.12 **Deforestation in Sumatra (after Whitmore)**

Primary forest and forest that is logged but not degraded

(a) Rainforest in 1932

(b) Rainforest in 1985

(c) The current human environment

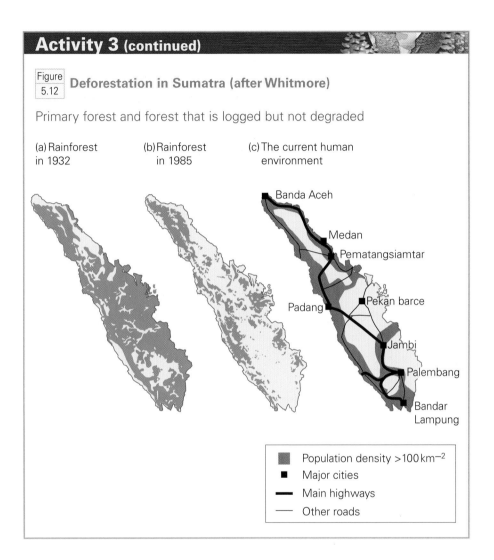

Banda Aceh
Medan
Pematangsiamtar
Padang
Pekan barce
Jambi
Palembang
Bandar Lampung

■ Population density >100 km^{-2}
■ Major cities
━ Main highways
─ Other roads

6 Temperate and polar biomes

Temperate deciduous forests: remnants of long-term clearance

Physiognomy and structure

Temperate deciduous forests cover 7 million km² in western and central Europe, eastern North America, eastern Asia, southern South America, southeast Australia and New Zealand (Figure 6.1).

| Figure 6.1 | **Global distribution of temperate deciduous forests** |

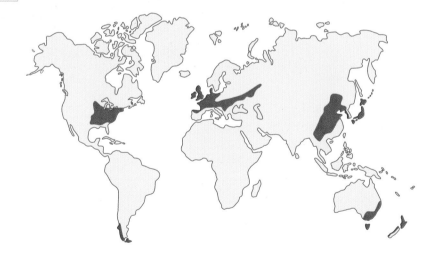

Climate varies significantly, with mean annual temperatures ranging from 5°C to 20°C, mean January temperatures from +3°C to −10°C and annual precipitation from 50 cm to 200 cm. In the southern hemisphere, winters are milder and the trees are evergreen (southern beech in New Zealand, myrtle beech and sassafras in Tasmania, eucalyptus in Australia). In the northern hemisphere, the trees are winter deciduous.

Forest structure is similar in America, Europe and Asia. The canopy is 20–30m high, consisting mainly of oak, maple, lime, beech, hickory, elm and ash. Beneath is a layer of subdominant, shade-tolerant trees, for example hazel, hawthorn, yew, sloe and holly, and shrubs such as rose and bramble. There is a ground layer of tree and shrub seedlings, herbs and ferns (Figure 6.2).

| Figure 6.2 | **Deciduous woodland of sessile oak, West Yorkshire** |

Michael Raw

When trees are leafless, 70% of the light reaches the forest floor. Some woodland flowers, such as bluebells, lesser celandines and orchids, take advantage of this by flowering in spring. In summer, 10% of the light reaches the ground, which is more than in tropical forests, and the understorey vegetation can be dense. Plant biodiversity is particularly high in North America, with 180 species of tree and 2000 species of flowering plant in southeastern USA. Biodiversity in Europe is lower, but still high.

Nutrient cycling

Litter-fall is important in the ecology of deciduous forests. It averages about 6 tonnes ha^{-1} yr^{-1}, of which 50% is autumn leaf-fall. This gives a thick mull humus that is important for nutrient cycling — it provides 80% of the nitrogen and phosphorus returned to the soil, 70% of the calcium and 40% of the potassium. These nutrients are released from the humus for plant uptake by the process of mineralisation by soil macroorganisms (e.g. earthworms, nematodes, mites, springtails) and microorganisms (e.g. fungi, bacteria). Although little natural deciduous forest remains, much is known about its productivity, food webs and nutrient cycling because of long-term experiments in ancient woodlands, such as Wytham Woods, Oxfordshire and Hubbard Brook Forest, New Hampshire, USA.

Post-glacial colonisation of the British Isles

Biogeographers use fossils, radiocarbon dating and pollen analysis to construct a picture of past vegetation. Wood remains, fruits, seeds, pollen grains and fern/moss spores are preserved in peat and lake-infillings. Figure 6.3 shows a pollen diagram of Holkham Mere, Norfolk prepared by Professor Harry Godwin, a pioneer of pollen analysis. The vertical sequence of pollen and spores in samples taken from this peat-core covers 10 000 years.

Figure 6.3   **A pollen diagram of Holkham Mere, Norfolk**

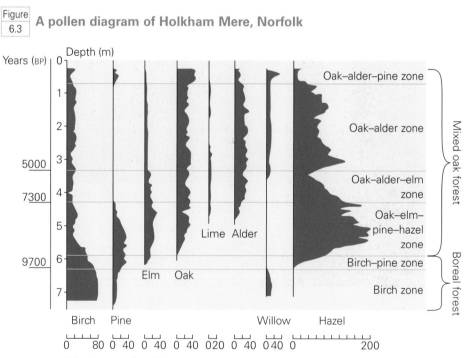

Pollen (except hazel) is expressed as a percentage of the total tree pollen.
The hazel scale relates to pollen count per unit of peat.
Age is shown on the left; equivalent modern forest types are on the right.

Using many pollen diagrams from all regions in the country, biogeographers have been able to piece-together the story of how plants in Britain have changed over recent geological time. In the Tertiary period, the trees were palms and tropical species. At the beginning of the Pleistocene Ice Age, oak, alder, hemlock, pine and spruce were common in western Europe, but were replaced by grasses, sedges, dwarf birch and willows during the coldest parts of the Pleistocene.

After the last ice age, trees migrated quickly back into the British Isles (Figure 6.3). The beginning of the post-glacial climate (**Holocene**) occurred about 10 000 years ago. Birch and pine were the first arrivals and the major deciduous trees responded quickly to the warming conditions; oak and elm were established by 9000 BP and had spread throughout the British Isles within 2000 years. Alder arrived by 8000 BP, lime by 7500 BP and ash by 6500 BP. Beech arrived about 3000 years ago, and reached only southern England and Wales before its range was altered by human activity (Figure 6.4).

| Figure 6.4 | **Beech woods** |

By 6000 BP, a climax deciduous forest, the **wildwood**, had developed everywhere, apart from high mountains, the far north of Scotland, and coastal dunes and salt marshes. Originally, it was called 'oak woodland', with oak thought to be the dominant species. However, studies by Rackham and Peterken show that numbers of lime, ash and hazel have been underestimated because they produce less pollen. They think that the forest consisted mainly of the shade-tolerant species, lime, elm and beech. As the forest was 'opened-up' by human activities, oak, which is less shade tolerant, replaced lime and elm. Figure 6.5 shows the five provinces of wildwood in the British Isles at 6500 BP.

| Figure |
| 6.5 |
Ancient native woodland at 6500 BP

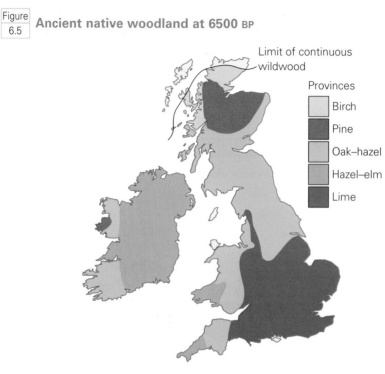

Limit of continuous
wildwood

Provinces

Birch

Pine

Oak–hazel

Hazel–elm

Lime

Forest clearance

Because of their favourable climates and fertile soils, most of the world's temperate deciduous forests have a long history of management and of being cleared for agriculture and settlement. Typically, only 1–2% of the original forest remains. These remnants are called **old growth forests** in North America and **ancient woodlands** in Britain (Figure 6.6).

| Figure |
| 6.6 |
 Ancient native deciduous woodland with bluebells, in Cumbria, early May

Michael Raw

Even these remnants have had some human use and impact; they are regarded as 'semi-natural' at best. Twentieth-century air pollution has caused further losses because acid rain attacks leaves directly and acidifies soils indirectly, which causes nutrient problems for plants. The present landscape of scattered forest 'islands' has lower biodiversity than the original wildwood — the smaller the forest fragment, the fewer the plant, bird and other animal species. Priorities in countries that have lost most of their temperate deciduous forests are:

- to preserve any remaining ancient woodlands in parks and nature conservation areas
- to re-establish woodland by offering financial incentives to farmers, and other landowners, to plant deciduous trees

Activity 1

Figure 6.7 **Structure of a temperate deciduous forest ecosystem**

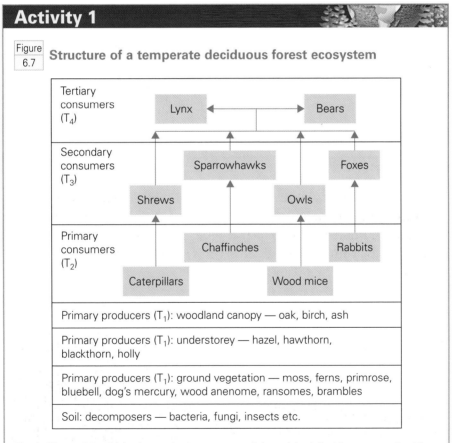

Study Figure 6.7, which shows the ecosystem of the original deciduous woodland in the British Isles.

(a) Describe and explain the vertical layering of plants in the woodland.

(b) Explain why many herbs in the deciduous woodland flower in spring.

Activity 2

The table below shows the results of a random sample of tree species in two ancient woodlands in northern England.

Tree species	Wood on sandstone, West Yorkshire	Wood on limestone, Cumbria
Oak	162	2
Sycamore	24	4
Birch	10	44
Beech	10	2
Ash	4	28
Elm	2	2
Yew	1	124

Investigate the influences of geology and soil on tree species of the two woodlands.

(a) State a suitable hypothesis to test the effect of sandstone and limestone on species composition.

(b) Explain the reasoning behind your hypothesis.

(c) Use the chi-squared (χ^2) statistical test to determine the significance of the difference in species composition.

(d) State and explain two factors, other than geology and soil, that may have contributed towards the difference in species composition.

Boreal subarctic forests: nature, exploitation and sustainable use

Distribution

Coniferous boreal forests (**taiga**) are found only in the northern hemisphere. The biome stretches from Alaska to Newfoundland in North America, and from Norway to the Pacific coast of Russia in Eurasia. They also occur in Scotland, in northern Japan and in mountains such as the Rockies, the Appalachians, the Alps and the Himalayas. Covering 12 million km², boreal forest comprises a quarter of the Earth's forest area (Figure. 6.8).

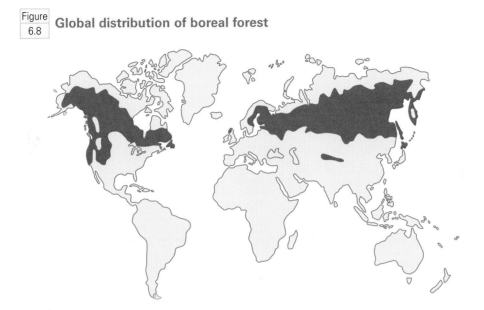

Figure
6.8 **Global distribution of boreal forest**

Its climate is subarctic, with mean annual temperatures between +5°C and −5°C. In winter, the minimum temperature often falls below −20°C. Summers are short and mild. Annual precipitation, with much falling as snow, varies between 20 and 200 cm, depending on the continent.

Physiognomy and structure

The needle shape of the leaves is a response to physiological drought in spring, when air temperature rises and trees start to photosynthesise, but moisture in the soil is still frozen. The evergreen habit allows the trees to take advantage of the short summer. In North America and Europe, evergreen species of spruce, pine and fir are dominant. However, in Asian Russia, the deciduous conifer larch is common — shedding the needles in winter is a response to the danger of frost damage and desiccation in the harsh Siberian winter. Further south, the taiga in eastern North America, western Europe and eastern Siberia merges into deciduous forest. In western North America and central Siberia, aspen parkland and temperate grassland lie to the south. Northwards, the boreal forest becomes open lichen woodland, in which scattered black spruce trees are separated by a striking ground cover of lichens.

The structure of the taiga biome is relatively simple. It is dominated by a few families of trees that have similar physiognomy and stand structure (Figure 6.9). The canopy layer of conifers is 20–30 m high and the height and density of trees decrease as temperatures become lower with increasing latitude and altitude.

Little light penetrates the closed coniferous canopy, so willow, alder and birch occur only at the edges of river valleys or in clearings. Some evergreen heathers and juniper are found, but there is little regeneration of dominant trees beneath their own canopies.

Figure 6.9 **Moose grazing in Canadian Taiga**

Ken Atkinson

Nutrient cycling

Soils are podzolic and, compared with those in deciduous forest, have a thicker litter layer of mor humus. This is because of lower rates of mineralisation by soil organisms, on account of the cold climate and acid soils. Therefore, cycling of nutrients is slow and net primary productivity is low. Like tropical forest trees (Chapter 5), some conifers in boreal forests have evolved mutualistic relationships with mycorrhizae. This guarantees nutrient absorption in these infertile, acidic soil conditions.

Risks of fire and insect attack

In Canada, about 2% of the biome burns each year, largely due to lightning fires. This produces fire-successions with soil moisture and fire frequency governing the types of trees. Dry sandy areas are fire-prone and burn at short intervals in their 'fire cycles'. This favours fast-growing species such as jack pine and lodgepole pine whose seed-cones need heat to open (**serotinous cones**) (Figure 6.10).

Figure 6.10 | Burned forest of lodgepole pine, Yellowstone National Park, USA

Michael Raw

Aspen and birch also sprout easily from roots and stumps in burnt areas. Moister areas have longer fire cycles. Hence, the conifers, for example fir and spruce, are more fire-sensitive. Where drainage is poor, with bog-peat and stagnant pools, black spruce grows in a landscape called **muskeg** by Native Americans (Figure 6.11).

Many diseases caused by insects and fungi occur in coniferous forests. Outbreaks cause defoliation and death in about 30% of trees. Thus, they are a severe problem for the forest industry. Most insect damage is caused by spruce budworm, pine budworm and spruce beetle, all of which occur in cycles of 15–30 years. It is estimated that each year, in Canada, 0.95 million hectares are lost to fire, 0.5 million hectares to insect damage and 0.75 million hectares are harvested. Logging is selective, whereas fire and disease affect trees of all ages. Losses from fire and insects account for 43% of the annual growth of forests, which is equal to the amount of timber removed.

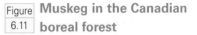

Figure 6.11 | Muskeg in the Canadian boreal forest

Ken Atkinson

Sustainability of forestry

Large parts of the coniferous forests in Canada and Asia are still wilderness, with minimal human impact from roads and settlements. Other regions, for example the Alps, have been used in a sustainable way for thousands of years for timber, firewood and forest-pasture. In recent years, the boreal forest has been the scene of disputes between developers, such as multinational forestry companies, and conservationists. For example, a bitter struggle continues between multinational forestry corporations and the non-governmental organisation Friends of Clayoquot Sound on Vancouver Island, British Columbia, Canada. In Siberia, there is concern that recent logging contracts given to Japanese and Chinese forestry companies allow harvests that exceed the forest's net primary productivity and are, therefore, unsustainable. Some people see the boreal forest biome as the major battleground in the future struggle to preserve global wilderness. The main human impacts on the boreal forest are summarised in Table 6.1.

Table 6.1 Human impacts on the boreal forest

Resource use	Impacts
Fuelwood	Variable sustainability; e.g. sustainable in the Alps but severe soil erosion caused by deforestation for fuel in Nepal
Logging	Extensive harvesting in Russia and Canada; clear-felling leads to devastated landscapes
Pulp and paper manufacture	Usually clear-felled; chemical pollution and toxic wastes
Pest control	Spraying of chemical insecticides
Recreation and sports	Hotels; hunting, skiing, snow-mobiles, snowboarding in winter; hunting, golf, horse-riding, camping and hiking in summer
Residential development	Prime residential sites; deforestation causing erosion
Mineral extraction	Metallic minerals cause chemical pollution when processed; hydrocarbon (coal, oil, natural gas) extraction results in wastes difficult to reclaim and land difficult to revegetate
Hydroelectric dams	Total loss of forest, the creation of ugly scars as a result of fluctuating reservoir levels
Farming	Subsistence agriculture for hay and livestock; clearance of southern margins for intensive forage more risky
Indigenous guiding, hunting and gathering	Provides food, clothing, artefacts and extra income; the ecosystem remains stable
Commercial sport hunting	Hunted species include deer, moose, bear, cougar, wolf, lynx and beaver; sound ecological management is necessary to maintain populations

In Scotland, the wildwood of the native Scots pine (Figure 6.12) has been reduced to small remnants in Speyside (Rothiemurchus, Abernethy and Glenmore forests), in Deeside (forests of Mar, Glentanar and Ballochbuie) and in other scattered patches, for example at Loch Rannoch and Loch Maree.

The Scots pine in these forests is a round-topped, open-branched tree, with the shape of a hardwood. It is very different from the straight 'pole' shape of pines in plantations (Figure 6.13).

Figure 6.12 **Native Scots pine woodland, Deeside, Scotland**

Ken Atkinson

Figure 6.13 **Plantation of Scots pine, Ballater, Scotland**

Ken Atkinson

After centuries of deforestation, there have been recent attempts at conservation. However, regeneration is hampered by two factors:

- suppression of fire (fire clears dense mats of heather and helps pine-seeds to germinate)
- grazing of tree seedlings by deer, sheep and rabbits

Tundra biomes: adaptations to cold and modern human impacts

Tree line and timber line

Tundra vegetation is found at high latitudes (arctic ecosystems) and high altitudes (alpine ecosystems). It occupies 8 million km² beyond the **tree line** (often called the **timber line** in mountains). The tree line is one of the Earth's major ecological boundaries. It defines the area where summer heat is no longer sufficient for tree seeds to ripen and for trees to produce enough net primary productivity for survival. It correlates with the 10°C July isotherm, which in turn marks the summer position of the arctic front separating cool arctic airmasses from milder maritime airmasses (Figure 6.14).

| Figure 6.14 | **Limits of the low Arctic and the high Arctic** |

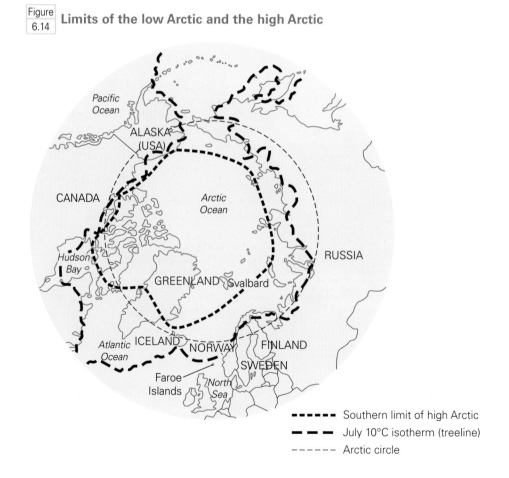

- - - - - - Southern limit of high Arctic
- - - July 10°C isotherm (treeline)
- - - - - Arctic circle

The presence of **permafrost** in the soil depresses soil temperatures and associated frostheave disturbs plants, producing a so-called **drunken forest** of dwarf trees. The growing season is short, usually less than 90 days, and, apart from in some coastal locations, precipitation is less than 10 cm.

Vegetation in the low Arctic and high Arctic

Tundra vegetation is simple in structure. Low plants with **prostrate habit** hug the ground. This means that they receive optimum warmth in summer and protection from wind beneath snow in winter. Vegetation zones change with increasing latitude and altitude, as the climate becomes more severe. Biogeographers recognise two main zones (Figure 6.15).

Figure 6.15 **Vegetation zones in the Subarctic and Arctic**

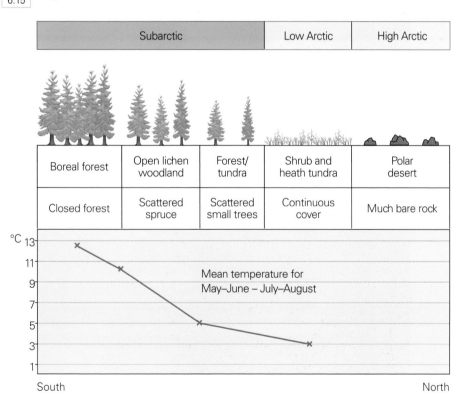

In the **low Arctic**, there is a continuous cover of ground vegetation, with sedges and mosses in wetter hollows, and heaths, grasses, lichens and herbs on drier ridges. Dwarf conifers, dwarf willow, dwarf birch and shrubs, such as alder, are found in sheltered sites (Figure 6.16).

Figure
6.16
Dwarf spruce in the low Arctic in Churchill, Canada

Ken Atkinson

At higher latitudes and altitudes (e.g. in the **high Arctic (polar desert)**) there is a predominance of bare ground. Lichens and mosses flourish but only a few flowering plants survive, in favoured patches. Two such species are the arctic poppy and the purple mountain saxifrage (Figure 6.17), the latter holding the world record for flowering the furthest north (Greenland, 83°N).

Figure
6.17
Purple mountain saxifrage in the polar desert of Devon Island in the Canadian high Arctic

Ken Atkinson

Nutrient cycling

Although the lack of heat is the major cause of the small biomass, low productivity and low biodiversity of arctic ecosystems, these also suffer from the low nutrient content of tundra soils. Only small amounts of organic matter are returned to the soil each year and the rate of decomposition is slow. There is also only minimal weathering of rock minerals in the soil to provide nutrients. Nitrogen levels are particularly low, so nitrogen fixation by algae and legumes such as *Oxytropis* is vital. The lack of soil nutrients is illustrated by lush growth where there is added nitrogen, phosphorus and calcium — for example, from the bones of dead animals and from droppings around birds' nests and animal burrows.

Adaptations to low temperatures

Tundra ecosystems exist because plants and animals have adapted to low temperatures, desiccating winds and snow-blast in winter blizzards. Plant adaptations that have evolved to aid survival in this inhospitable environment include:
- perennial habit
- prostrate form
- cushion form
- snow-hollow protection in winter
- vegetative reproduction
- pre-formed buds
- parabolic flowers
- heliotropism
- cell sap acting as 'antifreeze'

Animal adaptations include:
- migration
- hibernation
- thick fur, fat; feathers
- small extremities
- low surface area to volume ratio

Effects of acid rain and oil spills

Tundra vegetation exists in dynamic equilibrium with the underlying permafrost. During the short arctic summer, vegetation maintains lower soil temperatures because it acts as an insulator. Bare soil has a dark surface that absorbs more radiation, resulting in raised soil temperatures. Human activities remove vegetation directly for construction of buildings and transport, and indirectly by pollution. Toxic emissions of heavy metals and acid rain from metal smelters can devastate vegetation, as has happened in the Kola Peninsula, Russia. Permafrost melts in such situations; boggy scars with pools, slumps and gullies called **thermokarst** are formed. To avoid this ecological disaster, it is best to do as much transporting as possible in the winter, to build on thick gravel 'pads', and to use hovercrafts and balloon-tyred vehicles whenever possible.

Another serious impact is from oil spills in the arctic oilfields of Alaska and Russia. Figure 6.18 shows how crude oil impacts on tundra soils and vegetation.

Figure 6.18 **Impact of a crude oil spill on the arctic tundra, Alaska**

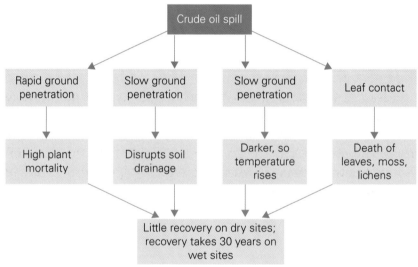

Dry sites, where the oil penetrates quickly, suffer most because the volatile chemicals in the oil kill roots. If the oil sits on the surface and does not penetrate, as in wet sites, the damage is less for the following reasons:

- Some toxic volatile substances in the oil evaporate.
- Although leaves are killed, the stems and roots survive and plants grow back in a few years.
- The darker surface causes an increase in soil temperature.

However, the effects of oil spills on mammals, birds and fish are severe.

7 Changes in ecosystems through time

Long-term changes over geological time

The idea that ecosystems change through time is not new. In the nineteenth century, Charles Darwin and Alfred Wallace recognised, by the nature and distribution of organisms over the millions of years of Earth history, that evolutionary change had taken place. In the past 150 years, increased knowledge of these changes has come from studies in palaeontology (rock fossils) and palaeoecology (plant remains in sediments and peat), and from a better understanding of the ecology and behaviour of contemporary species.

Continental drift

In 1915, the German meteorologist–geologist Alfred Wegener coined the term 'continental drift' to describe the movements of continents. Processes of plate tectonics, now known to drive these movements, were not discovered until the 1960s. About 250 million years ago, the world's continents were joined into the supercontinent Pangaea (Figure 7.1). About 200 million years ago Pangaea began to break into two landmasses — Laurasia in the north and Gondwanaland in the south. Thereafter, while animals and flowering plants were evolving, Laurasia broke up into North America, Greenland, Europe and most of Asia; Gondwanaland broke up into Antarctica, South America, Africa, Madagascar, Arabia, India and Australasia.

Figure 7.1 **Continental drift**

(a) Pangaea, 225 million years ago

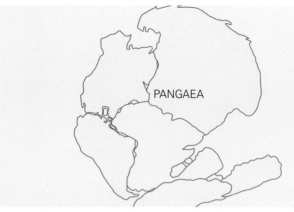

(b) Gondwanaland and Laurasia, 180 million years ago

(c) Many continents, 100 million years ago

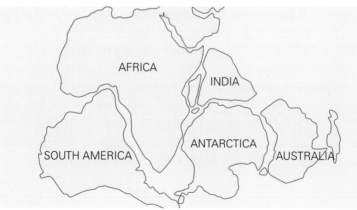

Some of today's plants and animals are single-species survivors of much more diverse groups of the past. They have been called 'living fossils', because they have not changed over long periods of geological time. Examples of living fossils include the maidenhair tree (gingko) and the metasequoia tree from China, the coelacanth (an Indian Ocean fish, rediscovered in 1938) and the tuatara reptile from New Zealand. Many organisms have a present-day range that can be explained only by all the southern continents having been joined together. Studies of the biology of species found in more than one southern continent, for example tree ferns (Figure 7.2) and velvet worms (Figure 7.3), indicate that divergence within the groups took place before the break-up of Gondwanaland.

Figure 7.2 **A tree fern**

Richard Smith

Tree ferns originated in Gondwanaland. Their range is now Australasia, southern Asia and Africa.

Figure 7.3 **Global distribution of velvet worms**

Dr Morley Read/SPL

Peripatidae
Peripatopsidae

The two families distinguished on the map evolved before the break-up of Gondwanaland

Changes driven by plant successions over short time periods

Succession and climax

In 1916, the American biogeographer Clements introduced the concept of **succession** or **seral change** to describe the processes of vegetation change that culminate in a stable, climax community. He regarded the 'climax' or 'equilibrium' as a state that could exist indefinitely and used the term **climatic climax** to indicate that this reflects the regional climate. In 1939, the British biogeographer Sir Arthur Tansley showed that in Britain, soil conditions also influence the natural wildwood; he preferred the term **edaphic climax**.

Plants colonise:

- bare, newly created land surfaces that result from volcanic activity, flooding, landslides or silt-deposition
- areas where vegetation has been removed or altered by a disturbance such as fire, avalanche, wind-throw or human action

The theory is that plant communities on all these surfaces develop in a predictable direction from lower plants, such as mosses and lichens, to grasses and herbs and, eventually, to more 'mature' vegetation such as forest.

- **Primary succession** occurs when a previously lifeless land surface is first colonised by plants and animals (e.g. the volcanic islands of Krakatau, Indonesia and Surtsey, Iceland).
- **Secondary succession** occurs when an existing ecosystem is recovering from a disturbance (Figure 7.4).

Figure 7.4 **A secondary succession in an abandoned field in the American midwest**

						Canopy
						Lower canopy trees
						Understorey trees
						Tall shrub understorey

| 1 yr | 10 yrs | 40 yrs | 100 yrs | >100 yrs |
| Annuals | Perennials | Shrubs | Young forest | Mature forest |

Time ⟶

Facilitation model

In the Clementsian model of succession, the seral stages are controlled by what he called **facilitation**. This means that the organisms that occur in the later seral stages depend upon changes to the habitat brought about by earlier plants and animals. Although the details of successional change differ with the particular type of succession, there are some general ecological processes at work in all successions (Table 7.1).

Table 7.1 Ecological processes at work in successions over time

Changes over time	Processes
Deepening of soil	Sedimentation and weathering of rocks
Stabilisation of soil	Binding by roots and soil humus
Growth of O and A surface humus soil horizons	Organic additions to surface and humification by increased populations of soil animals and microorganisms
Growth of E and B soil horizons	Leaching of chemicals (lime, salts, nutrients, iron) to form E, with deposition in B or transport into drainage
Increasing nutrient fertility	Greater rooting depth, and more active nutrient cycles
Better soil-water conditions	Thick soils avoid extremes of waterlogging and drought
Increasing biodiversity	More plant and animal species after the pioneer stage, but often decreases in the climax community
Decreasing soil erodibility	Better soil structures and increased binding by plants

Further study of succession has revealed that the pathways are more complex and unpredictable than in the simplistic, one-directional model of Clements. Other important factors need to be taken into account, including:

- the speed with which different species can colonise and grow
- the nearness of the site to a seed source, whether in the soil (the **seed bank**) or in the surrounding vegetation
- the inhibiting effects of some species on others
- random events

Hydroseres and haloseres

Hydroseres

A **hydrosere** is an ecological succession that begins in water or in a wet habitat and progresses through a characteristic zonation of vegetation to stable climax vegetation on dry sites. A **halosere** is a type of hydrosere that develops in salt water, whether inland or at coastal locations.

In a hydrosere that begins in fresh water, the succession is dominated by plants that are **hydrophytes**; succession in a halosere is dominated by **halophytes**.

Figure 7.5 shows the five ecological zones that are found in a transect from open water to dry land.

Figure 7.5 **Transect diagram showing zonation in a hydrosere**

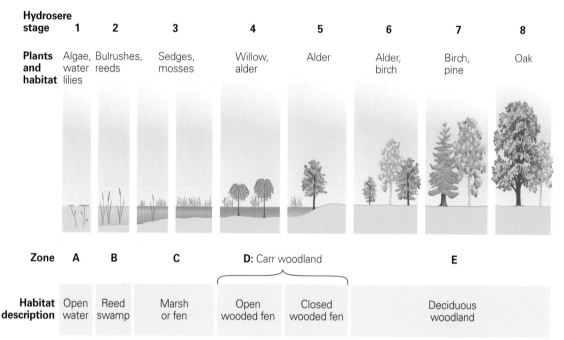

Hydrosere stage	1	2	3	4	5	6	7	8
Plants and habitat	Algae, water lilies	Bulrushes, reeds	Sedges, mosses	Willow, alder	Alder	Alder, birch	Birch, pine	Oak

Zone	A	B	C	D: Carr woodland		E
Habitat description	Open water	Reed swamp	Marsh or fen	Open wooded fen	Closed wooded fen	Deciduous woodland

Zone A is open water. As the depth of water decreases due to infilling by sediment and organic debris, plants such as water lilies, reeds and bulrushes are able to root in the mud. They trap additional sediment and produce more organic matter. Hydrophytes such as rushes and reeds can now colonise (zone B). Eventually, sediment rises to a point where it can be invaded by tussock sedges, hydrophytic grasses and mosses (zone C). Alder, willow and birch form a low, hydrophytic woodland, called **carr** (zone D). The site eventually becomes dry enough to support pine and a variety of deciduous trees (zone E). The role of alders in the succession is important because their roots have nodules containing microorganisms that fix atmospheric nitrogen into the nitrates needed by plants.

The precise plants found in the hydrosere depend on the climate and on the acidities of the lake and the soil water (Figure 7.6).

Figure
7.6 **Hydrosere in a dune-slack, Ainsdale National Nature Reserve, Merseyside**

Michael Raw

The water chemistry in turn depends upon geology and regional vegetation. For example, in the fens of East Anglia, mineral-rich water seeps through the peats and silts. There is high plant diversity, with many reeds, sedges and rushes around patches of water. On slightly higher sites, which remain above water in winter, alder, birch, willow, buckthorn and ash establish themselves, with well-developed shrub and herb layers.

Haloseres

Haloseres typically occur as **salt marshes** in low-energy environments along coasts — for example, in estuaries, behind spits or shingle ridges and between islands and the coast. They are also found in areas of inland drainage. At high tide, silt and organic matter are deposited from slack water and, over time, banks of mud are built up. The tides disperse the seeds of marsh plants and the plants further slow the movement of water, thus speeding up sedimentation.

Steers has studied the salt marshes of north Norfolk for a number of years. Rates of deposition of 2 cm yr^{-1} have been measured on the lower marshes of Scolt Head Island with lower rates recorded on older, higher marshes that are now covered only by high spring tides. On this coast, marshes form behind the protection of shingle ridges; they can develop in both seaward and landward directions (Figure 7.7). Reclaimed marshes are found behind the sea wall.

Figure 7.7 Topography of north Norfolk salt marshes, with shingle ridges, wild marsh and reclaimed marsh

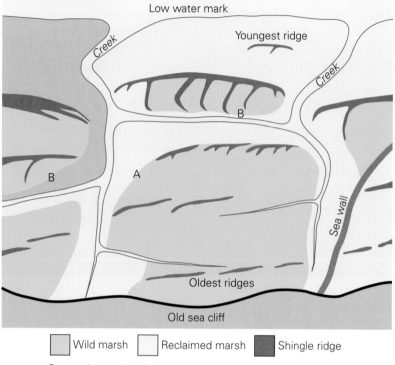

Low water mark

Youngest ridge

Creek

Creek

B

B

A

Sea wall

Oldest ridges

Old sea cliff

☐ Wild marsh ☐ Reclaimed marsh ■ Shingle ridge

Seaward growth at A; landward growth at B

Plant colonisation

Few plants can tolerate the harsh conditions of the salt-marsh environment; it is flooded regularly by seawater, which uproots seedlings and produces saline and anaerobic soils (Chapter 3). Plant communities are distributed over the marsh according to:

- the age of the marsh
- the height of the marsh
- the frequency of flooding (Figure 7.8)

Figure 7.8 Salt marsh on the River Kent estuary, Arnside, Cumbria

Michael Raw

The pioneer species on new mudflats is samphire, although in the past century this role has been partly taken over by a new invasive hybrid of cord grass. Low marshes are colonised by seablite, which prefers the higher and better-drained banks of creeks.

At higher levels, the upper-marsh community is sea lavender, sea pink, sea aster and salt-marsh grass. As the creeks become deeper, their higher banks are invaded by the dwarf shrub sea purslane. Above the mean high tidemark, soils are drier, less saline and flooded only by spring tides. The resulting vegetation is thicker and more diverse, with sea rush, grasses, sea plantain and sea wormwood joining sea lavender and sea pink.

Progressive leaching of salts over time reduces soil salinity, lowers soil pH and provides a better-drained surface. A range of grasses, herbs and shrubs are now able to colonise, eventually allowing pine and deciduous seedlings to invade.

Activity 1

Figure 7.9 **Cross-section through a salt marsh at Alnmouth, Northumberland**

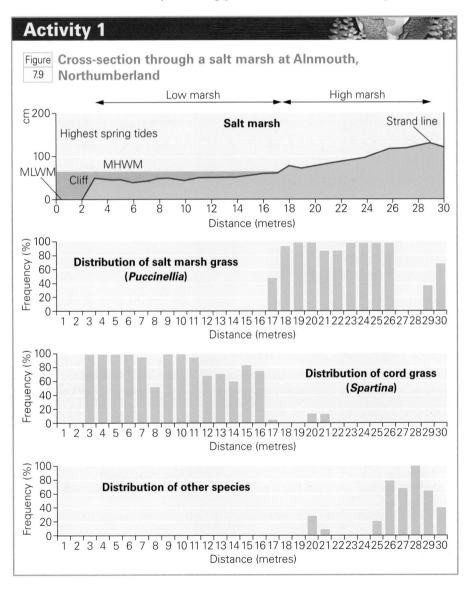

Activity 1 (continued)

Study Figure 7.9.

(a) Describe the contrasting plant distributions.

(b) Describe and explain why environmental conditions change from the mean low water mark (MLWM) to the strand line (height of the highest spring tides).

(c) Explain how these changes provide different opportunities for species to colonise and achieve dominance.

Xeroseres: psammoseres and lithoseres

Xeroseres are ecological successions where the main limitation to plant growth is drought. 'Psammos' (Greek, sand) and 'lithos' (Greek, rock) give the clues to these successions. A **psammosere** develops initially on sand; a **lithosphere** develops initially on rock. These initial sites hold little water for plant growth. Therefore, the facilitation processes in the succession provide colloids of humus and clay and a soil structure, all of which increase the **soil water-holding capacity** (Chapter 3). Coastal psammoseres start on the unstable beach, where blowing sand allows few plants to survive, exceptions being sea lyme grass and sea couch grass near the high tide mark.

Psammoseres

Marram grass

Marram grass is the most important builder of sand dunes. It is found on the seaward edge of the first unstable dunes and, thereafter, throughout the main dunes. Its importance cannot be overstated. It seldom reproduces from seed — instead it spreads by underground rhizomes. It is an aggressive coloniser and quickly stabilises the dune surface. As sand accumulates, the marram grows rapidly upwards, keeping pace with sand deposition. It dies out when the dune becomes stable and the deposition of new sand stops.

Once marram has provided stabilisation, a variety of plants, including sea sandwort, sea rocket, sand sedge, bird's-foot trefoil, thistles, yellow bedstraw, viper's bugloss, kidney vetch and ragwort, invade. Sandblast and salt spray harm coastal shrubs and trees, so these do not appear until large dunes can provide protection. Hawthorn is more tolerant and is an early coloniser. The scarcity of water in dunes is reflected in the xerophytic adaptations of the plants. For example, transpiration loss is reduced by:

- rolled leaves in marram
- hairy leaves in yellow horned poppy
- thick, waxy leaves in sea holly

Deep roots are also common.

Soil and microclimate

The psammosere provides a good example of how succession improves soil quality. With progressive plant colonisation, the initially inhospitable raw sand changes by four main facilitation processes:

- leaching of lime from shells, and salt from seawater from the dune-sand, which reduces pH from the initial value of about 8.5 to about 7
- the stabilisation of the mobile sand surface by the binding effects of plant roots and plant stems (sea bindweed, creeping thistle)
- the addition of plant residues to the soil surface, promoting humus formation, soil organism activity and nutrient cycling
- the formation of soil structures from the loose sand grains, giving enhanced water holding capacity and resistance to wind erosion

Two important factors in psammoseres are topography and microclimate. Rushes occur in damp depressions or 'slacks' within the dune system, wherever the water table reaches the surface. The crests of the dunes suffer from exposure to sandblast and salt, and from extreme windiness. Marram does not form a turf, so there are always some areas of bare sand for a storm to erode a 'blow-out' (Figure 7.10).

| Figure 7.10 | **The main sand-dune ridge at Sandscale, Duddon estuary, Cumbria, showing Marram grass and a 'blow-out'** |

Michael Raw

In conclusion, there are several successional pathways in the dune system, reflecting differences in initial conditions due to topography and drainage. These differing sites produce different 'climaxes' (Figure 7.11).

Figure 7.11 **Transect diagram showing the plant species in a coastal dune psammosere**

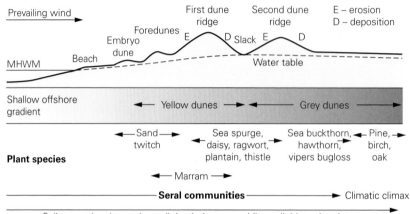

Activity 2

(a) Describe and explain how the facilitation model of plant succession works in sand dunes.

(b) Compare and contrast environmental conditions and vegetation on yellow dunes and grey dunes.

(c) What fieldwork methods, laboratory tests and statistical techniques might be used to analyse the relationships between soil properties and distance along a transect from the embryo dunes to the second dune ridge?

Lithoseres

Lithoseres develop more slowly than other successions because of the time required for the slow process of rock weathering to produce a mineral soil. The first colonisers are algae and lichens, followed by mosses, grasses and then low shrubs such as heathers and juniper. Variations in rates of formation and the nature of the succession depend upon the hardness and chemistry of the rock, and on the power of weathering and erosion processes at the particular site.

Disturbances

For a long time it has been recognised that human disruption by land management could produce 'altered' climaxes, called **plagioclimaxes** (or **disclimaxes** (*dis*turbed) in America). The grazing of grassland by livestock or the frequent burning of forest in shifting cultivation are examples (Chapter 5).

Important concepts in ecology, such as community structure, ecological succession and climax, are based upon the assumption that the physical habitat is stable in the long term. Accepting that disturbances are important, a modern definition of succession would be: 'the changes to the physical and biological conditions of an ecosystem that follow disturbances'.

Types of disturbance

In addition to human impacts, the importance of disturbances at work in the natural physical and ecological environment has been only recently appreciated. Disturbances may be powerful enough to prevent ecosystems reaching a stable climax. Disruption can be caused by various natural events, such as volcanic activity, fire, flood, landslide, avalanche, wind-throw, tsunami and attack by insects and fungi. For example, the Indian Ocean tsunami (2004) wiped out large areas of coastal marsh and mangrove forest, and Hurricane Katrina (2005) destroyed coastal and inland forests in Louisiana and Mississippi, USA.

Such disturbances are often described in the media as unnatural events or 'natural disasters', because of loss of life and the economic costs. However, they are natural and important events. Some ecosystems would be unable to sustain themselves without a disturbance. For example, in forests, shading by the canopy stops the growth of most plants, including seedlings of the climax trees. The death of trees creates small gaps, but regeneration of the climax trees requires large gaps to be made by major disturbances.

Fire

Wildfire has been well studied because it occurs in all biomes from the arctic tundra to tropical rainforests, and is probably the most frequent and widespread disturbance. It is costly because of the fire-fighting operations. It is a particularly powerful factor in coniferous forests, temperate grasslands, Mediterranean vegetation, savanna grasslands and scrub. The EU experiences 60 000 wildfires annually, which destroy 700 000 ha of forest and scrub, and cost 1.5 billion euros in fire fighting. Mediterranean ecosystems are particularly susceptible, with 500 000 ha of forest and scrub being burnt each year in the Mediterranean states of the EU (Portugal, Spain, France, Italy and Greece) and large areas of South Africa, California and south Australia also affected. The last 25 years have seen a tripling in the area burnt in this biome.

Fire susceptibility

Wildfires are influenced by three main factors: vegetation, topography and weather (Figure 7.12)

Figure 7.12 **Factors controlling wildfires**

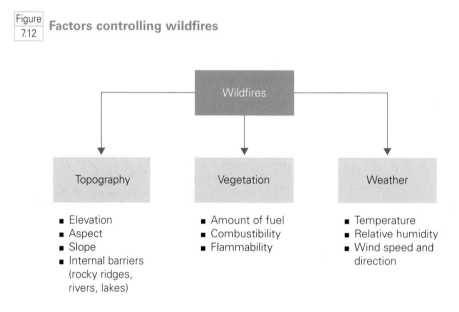

These factors can be mapped and modelled using modern GIS and remote-sensing techniques. The resultant models are used in countries such as Canada and Australia for fire prediction and control.

Pines and eucalyptus contain large amounts of resin and volatile oils — they are much more flammable than deciduous trees. Pines also retain lower dead limbs that can act as fire ladders. Lightning is the main source of natural fires, but even flammable pines are unlikely to ignite in a damp atmosphere. Fire susceptibility is increased by:

- flammable and closely spaced vegetation
- dry, hot weather
- topography without natural firebreaks, such as lakes, rivers and rocky ridges
- upslopes that increase the speed and intensity of fire
- dry winds that increase spread

Benefits of fire

Although portrayed in the media as destructive and negative, fire plays a beneficial ecological role in several ways. For example:

- Destruction of the canopy allows shade-intolerant plants to regenerate.
- Fires release mineral nutrients, such as potassium, calcium and magnesium, from dead and living vegetation back into the soil as ash. However, there is some loss of nitrogen and sulphur in smoke.
- Fires remove leaf litter and dead vegetation cover from ecosystems.

Some biomes depend on fire for regeneration. These include tropical and temperate grasslands (savannas and prairies), and forest ecosystems (Figure 7.13).

Figure 7.13 **Regeneration 9 years after a wildfire in Yellowstone National Park, USA**

Michael Raw

Plant adaptations to fire

There are a number of adaptations that help plants cope with the effects of burning (Table 7.2).

Table 7.2 Plant adaptations to fire

Adaptation	Plant examples
Fire-resistant bark	Douglas fir, ponderosa pine, western larch, cork oak
Deep root systems	Olive, grasses
Sprouting from branches and roots	Coastal redwood, eucalyptus
Sprouting from lignotubers between roots and shoots	Many eucalypts
Sprouting of suckers from root buds	Trembling aspen, paper birch, savanna and prairie grasses
Scarification — seeds dormant until heated	Banksia (Australia), lodgepole pine
Serotinous — fruits and cones will not open until heated	Jack pine, Aleppo pine

8 Conservation of biodiversity and wilderness

What is biodiversity?

Biodiversity is a shortened form of the term 'biological diversity' and is an important concept that was discussed at the Rio de Janeiro Earth Summit in 1992. It means the variety of plants, animals and microorganisms inhabiting a particular place on Earth: in other words, 'the totality of all ecosystems, species and genes in one location'. It is more than just a list of species, and includes:

- differences in genes within a species
- differences between species
- the health of all interactions and interconnections in the ecosystems that animals and plants inhabit

The concept overlaps with the concept of **ecological integrity**, used in conservation planning and management (p. 129).

Species richness

The total number of plants and animals on Earth is not known. So far, 1.7 million types of viruses, microorganisms, plants and animals have been identified, of which 56% are insects and 14% plants. The discovery of a new mammal, the Cyprus mouse, in Europe in 2006 illustrates how much there is still to learn about biodiversity on Earth. Most estimates of the number of species range from 4 million to 20 million. However, some people believe that there may be 200 million species of microorganism alone. Such data are examples of **species richness**. This is measured by counting the number of species in a given area and extrapolating.

The total number of plant species is thought to be between 270 000 and 422 000. It is an imprecise estimate because of lack of knowledge of many rare species, which make up most of all biological groups. Assuming 400 000 species of flowering plant on Earth, of these, the UK has 1500, compared with 6000 in Spain. Thus, Spain has four times the flowering plant species richness of the UK. There are a number of factors that explain this, one of which is area. Other things being equal, larger areas will have more species than smaller areas; the area of Spain is twice that of the UK.

Activity 1

Channel Island	Area (ha)	Species richness of higher plants
Jersey	11 808	766
Guernsey	6560	804
Alderney	814	519
Sark	525	425
Herm	131	255
Burhou	30.4	18
Jethou	18.1	186
Lihou	15.6	99
Crevichou	1.2	45

Study the table above.

(a) Draw a scatter graph of the data.

(b) Calculate the Spearman rank correlation coefficient. What does this tell you about the relationship between species richness and area?

(c) Calculate the coefficient of determination. What does this tell you about the relationship?

(d) In addition to area, what other factors might explain the variation in species richness?

Genetic biodiversity

Counting species is a much-used measure of biodiversity. Biogeographers are interested also in different **genotypes** of a single species. For example, the Scots pine and the grey wolf both have large ranges in Eurasia and North America, yet in each case populations in different regions are genetically different, although all are classed as the same species. Genetic analysis of the DNA of plants and animals is a new science that will eventually give information about the true biodiversity of the Earth.

Latitudinal gradients of biodiversity

A pattern with most biological groups is that there are more species in the tropics than in higher latitudes in the northern and southern hemispheres. Table 8.1 summarises data for selected plants, mammals and insects that illustrate this. Not only does the total number of species increase towards the equator, the average number of species per unit area also increases. There are a few exceptions — for example, whales and sandpipers. Some groups, for example pines and aphids, reach maximum diversity in middle latitudes.

Table 8.1 Latitudinal gradients of biodiversity

Group	Number of species	Location/habitat
Trees	9	Canadian boreal forest
	620	Total in USA
	4500	Total in India
	>10 000	Rainforest in tropical Latin America
Mammals	18	Northern Canada
	150	Central America
Ants	3	Alaska
	63	Utah, USA
	222	Brazil
Butterflies	134	Michigan, USA
	750	All North America
	1550	Panama
	7000	Central and tropical South America

Historical theories

Numerous explanations have been offered to explain latitudinal gradients of biodiversity. The historical theories point to alternating glacials and inter-glacials in the Pleistocene, with environmental disruption in middle and high latitudes wiping out species. In contrast, long periods of environmental stability in the tropics allow time for the evolution of species. This is the **stability–time**

hypothesis. It is now known that the tropics were cooler and drier in the Pleistocene, which in some areas resulted in increased rates of speciation. The fact that Lake Baikal in Russia has existed for 1 million years and has 580 species of deep-water invertebrates, while the Great Slave Lake in Canada, which is 10 000 years old, has only four species, fits this hypothesis.

Non-historical theories

Some biogeographers argue that modern conditions, rather than historical ones, explain biodiversity gradients. The **habitat–diversity hypothesis** emphasises the diversity of vegetation structure, topography, soil and microclimate. In tropical forests, vegetation stratification provides numerous niches for birds, insects, primates, amphibians and reptiles (Chapter 2). However, this hypothesis does not explain why biodiversity is higher in the tropics than in structurally similar regions in the middle latitudes — for example, tropical savanna has higher biodiversity than temperate prairie.

Competition, predation and productivity hypotheses

The **competition hypothesis** stresses how evolution in the tropics is driven by interspecific competition, leading to narrow specialised niches (Chapter 2). It works well for mammals, amphibians and reptiles, but not for trees.

The **predation hypothesis** argues that prey populations are kept so low by intense predation in tropical forests that one prey species never dominates, so more can evolve. This is the generalised case of Janzen's 'enemies hypothesis' — seed predation explains why most tropical trees occur as isolated individuals (Chapter 5).

The **productivity hypothesis** argues that high primary productivity provides more energy, which can support more species, especially at the higher trophic levels. The correlation works well at a global scale, but breaks down at smaller, regional scales. For example, ecosystems such as estuaries and marshes have the highest primary productivities in the world.

Conclusions on biodiversity gradients

History has a role to play, but does not explain entirely the differences in biodiversity at high and low latitudes. A clear relationship exists between climate, primary productivity and species richness. Using evapotranspiration rates as an index of available light, heat and moisture, Figure 8.1 shows that positive feedback relationships exist between these key resources and the diversity of trees and vertebrates in North America.

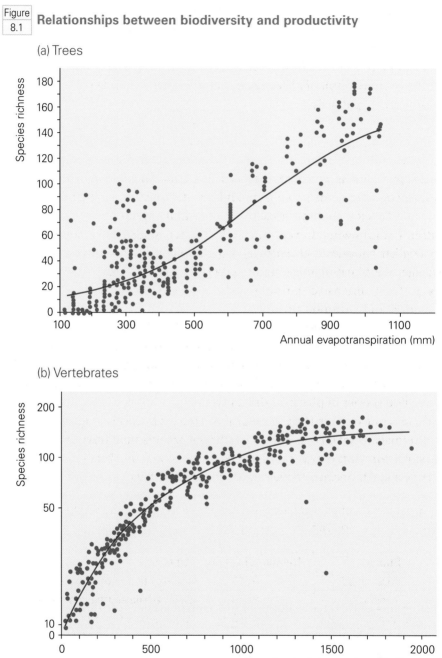

Figure
8.1 **Relationships between biodiversity and productivity**

(a) Trees

(b) Vertebrates

Furthermore, low latitudes have smaller seasonal differences in climate. Additional features are the complex vegetation structure, increasing niche specialisation and the abundance of prey.

In summary, biodiversity gradients on Earth do not have a single cause. They are the complex result of historical, ecological and environmental factors acting at large and small scales.

Losses of biodiversity

Much publicity is given in the scientific literature and in the media to pessimistic statements about the numbers of species in danger of extinction. In 1980, Lovejoy predicted that 10–20% of all living species would be extinct within the following 40 years. In 2000, Marren studied published floras for 16 English counties and discovered that, during the twentieth century, the average rate of extinction was one species per county per year. These losses result from intensive farming, urban development, wetland drainage, pollution and eutrophication.

The World Conservation Monitoring Centre based in Cambridge and at Kew Gardens collects data from all continents and publishes the *Red List of Threatened Plants* — species in danger of extinction. Table 8.2 shows the World Conservation Monitoring Centre categories and the number of species in each category. There are 33 000 species of plants 'at risk' or worse.

There is much uncertainty in predicting future extinctions of species of plants and animals. What is clear, however, is that present distributions of many species are small remnants of past distributions. Thus the wolf and bear in Europe today occupy a small fraction of their past range (Figure 8.2).

Table 8.2 Numbers of threatened plant species according to the World Conservation Monitoring Centre

Status	Number of species	Meaning
Endangered	6 500	Immediate danger of extinction
Vulnerable	8 000	Approaching endangered status
Rare	14 500	Exposed to abnormal risk
'Unknown status'	4 000	Unknown level of risk
Total	**33 000**	—

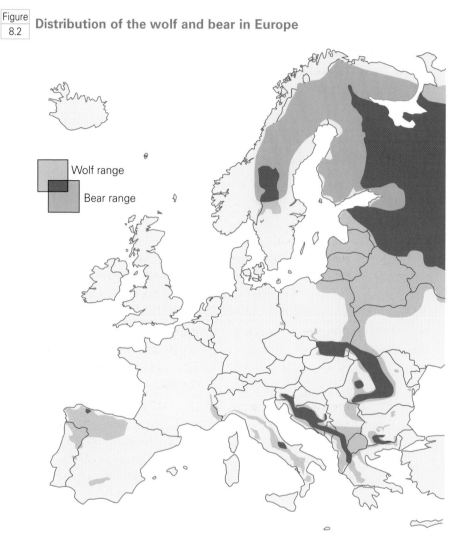

Figure 8.2 **Distribution of the wolf and bear in Europe**

Wolf range

Bear range

Conserving biodiversity

Maintaining and enhancing biodiversity are not simple matters and a number of strategies need to be followed. Inputs are required not only from bio-geographers, but also from biologists, information technologists and social scientists, including planners. Essentially, three approaches can be used to conserve species and their habitats:

- protection
- conservation planning, including gap analysis
- restoration

Protecting species and habitats

Protection by giving a **designated status** is used at all levels — international, national, regional and local. Table 8.3 illustrates some examples of the types of protected area at these different levels.

Table 8.3 Designations for protecting nature

Levels	Institutions	Protected areas
International	International Union for the Conservation of Nature (IUCN)	Category 1 Protected Areas
	UN Educational Scientific and Cultural Organisation (UNESCO)	World Heritage Sites of 'outstanding universal value', e.g. Studley Royal Park and Fountains Abbey in North Yorkshire, the Jurassic Coast in Dorset and Devon.
	European Union (EU)	Special Areas of Conservation (SACs) (EU Habitats Directive); Special Protection Areas (SPAs) (EU Birds Directive)
National	Departments for parks and nature (e.g. Countryside Agency in UK; Natural England; Scottish Natural Heritage)	National Parks — sites designated for aesthetics, recreation and tourism as well as wildlife; National Nature Reserves (NNRs) and Areas of Outstanding Natural Beauty (AONBs) in the UK; National (Federal) Wildlife Reserves in USA; 'Country' areas in Canada
	NGOs	RSPB, Woodland Trust, Wetland and Wildfowl Trust reserves
Regional/ provincial/ state	Governments of regions, provinces and states	Regional Parks (UK); Provincial Parks (Canada); State Parks and Wildlife Reserves (USA)
	NGOs	Nature reserves of the County Wildlife Trusts, e.g. London WT, Norfolk WT
Local/city/ municipal	Local government	Local Nature Reserves (LNRs); city wildlife reserves; Nose Hill Wilderness Area, Calgary, Canada

Conservation planning

Protected areas usually have planning restrictions to prevent habitat loss. Protection can range from the strong and absolute, as in nature reserves and national parks (Figure 8.3) in many countries (but not in the UK), to weak protection only, as in Areas of Outstanding Natural Beauty in the UK (Figure 8.4) and 'Country' designations in Canada (e.g. Kananaskis Country near Calgary).

Figure 8.3 Arches National Park, Utah

Figure 8.4 Cross Fell, a National Nature Reserve and Area of Outstanding Natural Beauty, northern Pennines

Sites such as World Heritage Sites and National Parks act as magnets for visitors and this increased tourism brings extra pressure on wildlife.

In the UK, Biodiversity Action Plans have become important for conservation planning at both local and national levels. Although much used, they are controversial. Marren thinks that they put too much emphasis on using scarce resources to maintain the status quo and in protecting rare species from natural processes — for example, using imported sheep to graze a meadow to preserve a rare orchid (Figure 8.5). This 'interventionist' approach is not a long-term solution; there is a case for conservation goals involving less management and more natural successions.

Figure 8.5 **Military orchid: a plant rarity found in only two nature reserves in the UK**

Michael Raw

Gap analysis

Gap analysis uses a similar approach to protect biodiversity. It uses GIS (geographic information systems) to identify ecosystems and species that are not protected adequately in conservation programmes. The technique uses overlays of several data sets of plant and animal distributions (particularly rare species) and the distribution of currently protected areas to produce a data set and map of gaps that are used to identify species not represented in protected areas. Although it is possible to interpret overlay maps visually, gap analysis is much more powerful using computerised GIS because it can hold many more data sets and model more easily the complexity of real ecosystems.

Restoration of species and habitats: 'rewilding'

Rewilding schemes are designed to re-establish lost habitats and to reintroduce lost species. Intensive farming is judged to be a major reason for loss of habitat and species, so agri-environment schemes, for example Environmentally Sensitive Areas and Countryside Stewardship, may stabilise the losses. The reintroduction of species that have become locally extinct has met with success. For example, the capercaillie became extinct in Scottish pinewoods in the 1770s, but was reintroduced 70 years later; the sea eagle was reintroduced into northwest Scotland in 1986 and the red kite into northern England in 2000.

An initiative is underway in Scotland to re-establish the beaver and there is talk by enthusiasts of re-establishing wolf, bear, moose and lynx, which used to live in the Caledonian pine forest (Figure 8.6).

An example of an ongoing rewilding programme is the reintroduction, in 2005, of bison into the Grasslands National Park, Saskatchewan, Canada, after an absence of 130 years (Figure 8.7). Although wolf and bear would have been part of the pristine prairie ecosystem, their introduction has not been proposed because of the sensitivity of local ranchers to the idea.

Figure 8.6 Caledonian pine forest, Glentanar, Scotland

Ken Atkinson

Figure 8.7 Bison being reintroduced into the Canadian prairies in the Grasslands National Park, Saskatchewan

Pat Fargey

Restoring native woodland is an active field in the conservation of biodiversity. It involves removing non-native trees and excluding deer, so that native trees can regenerate. Table 8.4 lists some projects that are currently in operation in the UK.

| Table 8.4 | Restoration of woodland in Britain |

Woodland project	Manager
Carrifan Wildwood, Ettrick Forest	Borders Forest Trust, Scotland
Cashel Farm, Loch Lomond	Royal Scottish Forestry Society
Dartmoor	Moor Trees
Glen Affric, Scottish Highlands	Trees for Life
Glen Finglas, the Trossachs	Woodland Trust
Mar Lodge, Cairngorms	National Trust for Scotland
Millennium Forest of the Yorkshire Dales	Yorkshire Dales National Park
Millennium Forest of Scotland	'Reforesting Scotland'
West Affric Estate, Scottish Highlands	National Trust for Scotland

Wilderness as natural and cultural space

Definitions of wilderness

Early definitions

It is not easy to define 'wild' and 'wilderness' because their meanings are partly subjective. As a result of personality and experience, people have different perceptions and expectations. There is a saying: 'one man's wilderness is another's roadside picnic ground'! The meaning of wilderness has changed over time. It was defined by the citizens of ancient Greece as 'waste ground' — the space furthest away from the civilising space of the city-state itself. In the Bible, it is a harsh wasteland, where the faithful go for spiritual renewal. The Anglo-Saxon derivations are: *'wild*-doer'—a wild beast and 'wild-doer-ness' — the place of wild beasts.

Modern definitions

The USA was the first country to give a legal definition in its Wilderness Act of 1964: 'an area where the earth and its community are untrammelled by man, where man himself is a visitor who does not remain'. This definition owes much to the vision of the Scottish conservationist John Muir, who emigrated to the USA in the mid-nineteenth century and valued the wildness of North American landscapes in contrast with the tamed, cultural landscapes in overcrowded Europe.

The American concept ignores aboriginal peoples, who, since prehistoric times, have used all the wildernesses on Earth (apart from Antarctica) for hunting, gathering and herding animals. Other definitions recognise this. In Canada, there is pride in traditional aboriginal homelands as shown by campaigns for 'land claims' and 'aboriginal rights' by native groups such as the Inuit in the tundra, the Cree in the boreal forest and the Blood Indians in the prairies. Here wilderness is defined as: 'an area where natural processes dominate, yet where people also co-exist, as long as their technology and impacts do not endure' (Figure 8.8).

| Figure 8.8 | **Aboriginal people in the wilderness of the Canadian Arctic, Baffin Island** |

Ken Atkinson

In Europe and in countries long-settled by Europeans, the definition is usually amplified to include traditional and sustainable multiple uses by both aboriginals and Europeans. For example, according to Finland's 1991 Wilderness Act the definition is: 'a culturally and historically defined term. It used to be part of people's everyday life as hunting and fishing areas'. In the 1940s, the American conservationist Aldo Leopold viewed wilderness as: 'a continuous stretch of country, preserved in its natural state, open to lawful hunting and fishing, devoid of roads, artificial trails, cottages and other works of man'.

Anthropocentric and biocentric wilderness

There are two conceptions of wilderness:
- **anthropocentric** — judges that human needs are paramount and stresses the recreational value of wilderness
- **biocentric** — defines wilderness in ecological terms and judges quality by lack of human disturbance

Central to the biocentric ethic is the belief that the Earth's wild ecosystems are intrinsically valuable and attractive — not just because we can use them. While this view has become increasingly important in recent years, the anthropocentric view remains dominant. The human use of wilderness continues to be a critical factor in arguments in support of policies for its preservation and management (pp. 132–133).

Wilderness in Britain

Some people argue that true wilderness no longer exists in Britain. This is true in comparison with the pristine wildland found in areas such as Alaska, Canada or Antarctica. Thousands of years of settlement, agriculture, forestry, transport and industry have produced almost entirely humanised landscapes. However, there is wildland in the Highlands and islands of Scotland, the southern uplands, the Lake District, Wales, the Pennines, the North York Moors, Dartmoor and Exmoor.

These areas are as much 'cultural' as 'natural' wildlands. They have been influenced by millennia of human impacts, starting with forest clearance in Mesolithic and Neolithic times, and followed by grazing by sheep and cattle, mining, planting of conifers, burning of heather for grouse shooting, and management of deer-forest. Yet they possess significant elements of wildness and have a perceived naturalness. While not ecologically pristine, by British standards they are remote, inaccessible areas of solitude with rugged terrain and scenic grandeur. They present physical challenges and risk, particularly to urban dwellers wishing to experience nature in a more natural setting.

Wildland in Scotland

Scottish Natural Heritage and the National Trust for Scotland avoid using the term 'wilderness'. They prefer **'wildland'**, which the National Trust for Scotland defines as: 'relatively remote and inaccessible, not noticeably affected by contemporary human activity, and offering high-quality opportunities to escape from the pressures of everyday living and to find physical and spiritual refreshment'.

Perhaps the discussion of wilderness in Britain needs to put less emphasis on ecological integrity and to focus more on perceptual values. **Ecological integrity** is a term much used in planning national parks, for example in Canada. It refers to the quality of the ecosystem. It is a measure of how healthy the natural ecological components and processes are and how well they are working together.

Scotland and Scandinavia are often quoted as having the most wildland in Europe. However, this is debatable as there are significant wildernesses in Russia, Poland and Spain.

The value of wilderness

Five values of wilderness

Five types of value come from wilderness:
- **experiential** values of the 'wilderness experience' in recreation and tourism, including aesthetic appreciation, spirituality and escapism
- **mental renewal** brought by communing with nature and solitude in a stress-free environment
- **scientific values** for research, education and discovery, whether experienced personally or enjoyed through books and the media
- less tangible values — wilderness as an **inspirational muse**. In 1851, the American conservationist Thoreau wrote: 'in wildness is the preservation of the world'. This is a reminder that there is a spiritual dimension to wilderness; qualities judged to be 'awe-inspiring', 'desolate', 'forbidding', and 'magnificent' provide inspiration for artists, philosophers, photographers, poets, musicians, theologians and writers.
- **economic values** (explored below)

Economic values

The economic values come from recreation and tourism, and from the value of genetic resources from plants and animals living in natural ecosystems. Using statistics from *Visit Scotland* and Scottish National Heritage, the following conclusions can be drawn:
- Tourism accounts for 7% of employment in the UK as a whole and 18% in the Highlands and islands of Scotland.
- Tourism is worth over £4 billion annually to the Scottish economy.
- 'Beautiful scenery' is the most commonly cited reason for people visiting Scotland; 90% say it is one of the reasons.
- Of domestic holiday trips to Scotland, 30% include activities such as hiking or hill-walking.
- Two per cent of the population of Great Britain visit the Highlands and islands for mountaineering activities (Figure 8.9). This is half of the 4% of the British population who mountaineer.
- Mountaineers provide 25% of the total tourist spend in the Highlands and islands.

Figure 8.9 **Mountain sports on Cairngorm summit, Scotland**

Steve Carver

Recreational use of wilderness

The remote and semi-natural wild places in Britain attract urban dwellers seeking a 'wilderness experience'. Table 8.5 lists activities that are appropriate because they are in keeping with wilderness values, particularly the maintenance of ecological integrity.

Table 8.5 Appropriate uses of wilderness areas

Uses	Types
Backpacking	Casual, expeditions
Camping	Primitive
Cycling	Trail biking
Education	Public media, school and university studies
Heritage appreciation	Art, culture, photography
Hiking/walking	Dispersed, guided tours, interpretive trails
Horseback riding	Casual, guided, expeditions
Interpretative programmes	Audio-visual, demonstrations, school activities, special events
Nature appreciation	Bird-watching, plant and wildlife observation
Orienteering	Exploration
Photography	Commercial filming, personal
Picnicking	
Scientific research	
Sightseeing	Plants, scenic vistas, wildlife
Wilderness skills	Survival skills, communing with landscape and wildlife

Managing wilderness

Wilderness management needs to be comprehensive to be effective. It involves work by scientists, for example biogeographers, in evaluating and identifying wilderness and in guiding decision making by planners and politicians, who need to balance public views. Table 8.6 identifies the main activities involved.

Table 8.6 Managing wilderness

Principle	Action
Setting goals and ideals	Education, consultation, decision making
Collecting scientific knowledge	Identification, research, surveys
Making policy decisions	Publication, public involvement
Managing	Seven pillars of wilderness management (p. 133)
Engaging the public	Education, consultation, decision making

The wilderness continuum

In a number of countries there are programmes for identifying and mapping wilderness. It is usual to recognise wilderness as one extreme in a continuum of human modification of the environment. There is a sliding scale from 100% built environment in the city centre to pristine nature in remote locations. Remoteness and naturalness are the two key indicators in this **wilderness continuum** (Figure 8.10).

Figure 8.10 The wilderness continuum

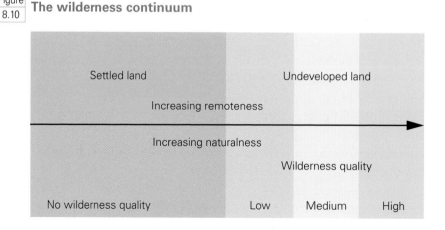

Carver produces maps of wilderness quality in Britain using the wilderness continuum approach (Figure 8.11). His geographic information system (GIS) uses 10 data sets:

- protected areas
- altitude
- naturalness of land-use
- apparent naturalness
- population access

- population density
- distance to settlements
- rail accessibility
- road accessibility
- transportation access

Figure 8.11   **Wildland continuum of Scotland based on 'naturalness'**

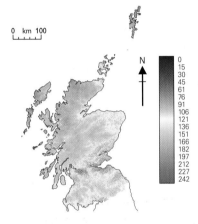

0   km   100

N

Example wildland continuum
Based on a combination of standardised maps stressing naturalness (standardised scale)

0
15
30
45
61
76
91
106
121
136
151
166
182
197
212
227
242

Seven pillars of wilderness management

Managing the Earth's remaining wilderness is a challenge for scientists and public alike. There is much to be learnt, yet little time to make conservation planning effective. The 'seven pillars' that should guide decision making in managing these valuable areas are as follows:

- Respect overriding legislation and land-use plans.
- Allow natural forces to shape the landscape.
- Use minimal management tools.
- Favour wild indigenous species, especially wilderness-dependent endangered ones.
- Encourage hunting, trapping and angling if compatible with the wilderness experience.
- Promote wilderness research into ecosystems, wildlife and people-in-wilderness.
- Be comprehensive in management, covering all aspects of natural complexity.